OXFORD
POLYTECHNIC

A GUIDE TO FOOD ADDITIVES AND CONTAMINANTS

A
GUIDE
TO
FOOD ADDITIVES
AND
CONTAMINANTS

By

K. T. H. FARRER

The Parthenon Publishing Group
International Publishers in Science & Technology

Casterton Hall, Carnforth,
Lancs, LA6 2LA, U.K.

120 Mill Road, Park Ridge
New Jersey, U.S.A.

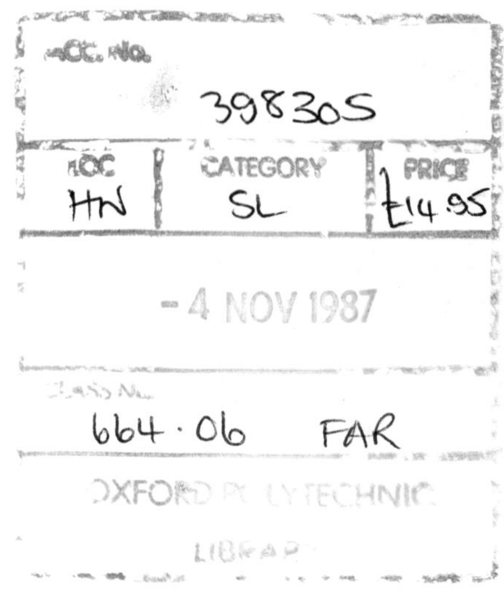

Published in the UK by
The Parthenon Publishing Group Limited
Casterton Hall, Carnforth,
Lancs, LA6 2LA, England
ISBN: 1-85070-127-X

Published in the USA by
The Parthenon Publishing Group Inc.
120 Mill Road,
Park Ridge,
New Jersey 07656, USA
ISBN: 0-940813-11-4

Printed and bound in Great Britain
by Antony Rowe Limited, Chippenham, Wilts

Contents

Foreword

The unrecognized triumph of industrialized societies is the year-round availability of an exceedingly varied and safe diet of high nutritional quality. The achievement has been gradual and few people under the age of fifty can imagine what food was like before food science and technology had made the modern scene possible. As you next look at the wealth on the supermarket shelves, think please, of the healthy (?) country fare on which my parents were raised and millions more like them a hundred years ago.

Milk was uncontrolled and was an excellent vehicle for the tuberculosis which was then rife, animals were slaughtered without regard to the elementary rules of hygiene, vegetables and fruit were highly seasonal and often pest ridden, fish was stale and expensive unless caught locally. Bread, potatoes and oatmeal were the dietary staples. Chemical preservatives, some of a much more dangerous kind than those in use today, were widely used and if food was not contaminated by pesticides it was often contaminated by pests. Legal protection of the public from gross adulteration of food was in its infancy. The domestic refrigerator had yet to be invented. In a word, food was often dull, always inconvenient, and sometimes downright dangerous.

But the present generation of shoppers knows nothing of this. It does know, however, that science is widely applied in producing the familiar shelf-wealth that goes into the shopping trolley. There is a feeling abroad that modern food is not 'natural', and only natural food is good. Manufacturers and food scientists jazz up impoverished products with chemicals to make them look and taste good and to hell with safety and nutrition.

This makes a good media story and the truth is not in it. The misguided or malicious have made unimpaired play with it. Balanced judgements, the most reliable guide available to us, are replaced by speculation and innuendo. 'We must have evidence that our food is absolutely safe' is the cry, but it is impossible to prove that anything is absolutely safe. One can but minimize risks.

This is what this book is about. It balances risks with benefits. It gives the facts. To read it is to be persuaded that we are fortunate indeed to enjoy the safety and luxury of our modern food supply. But I write of the western countries. Only those who have travelled in the poorer and hungrier parts of the Third World can understand just how fortunate we are.

31st March, 1987

John Hawthorn
Former President of the International Union of Food Science and Technology and Professor Emeritus of Food Science, the University of Strathclyde

Preface

> The days of the gross adulteration of food have probably passed into history but the timid householder is frequently perturbed by the discovery that things are not always what they seem, even under the most innocent and attractive labels.
>
> *Argus*, Melbourne, 14 February 1903

As a contribution to a more balanced understanding of the vexed question of food additives and contaminants, this book was first published in Australia by the Melbourne University Press under the title of *Fancy Eating That!* It was written originally for an Australian and New Zealand readership but in preparing it for a wider public each chapter has been modified and Chapter 7 has been re-written.

A *food additive* is a substance deliberately added to a food by the manufacturer to facilitate processing or to improve appearance, texture, flavour, keeping quality or nutritional value. On the other hand, a *chemical contaminant* in a food is a substance which is not normally present in that food in its natural form, or which is present in concentrations not normally found, or which is not permitted under the food regulations to be present or, being an additive as defined under the regulations, exceeds the concentration permitted. *Microbial contaminants* are much less easily defined. In its simplest form, microbial contamination is the presence in the food of unwanted microbes, but considerations of numbers, growth, spoilage and disease also arise. Additives have their place; contaminants do not. All are under continuous study by regulatory authorities all round the world, but there is need for vigilance lest unexpected misfortunes such as the thalidomide disaster in the drug field or the mass poisoning by contaminant mercury at Minimata in Japan overtake us.

I am particularly indebted to Mr M. P. Jackson, a former chairman of the Food Science and Technology Sub-committee of the National Health and Medical Research Council, Canberra, and Mr J. S. Fraser, Assistant Director (Food Standards), Department of Health, Wellington, for willingly supplying background information, and to my friend and former colleague, Miss Margaret Dick, whose comments on the draft of Chapter 6 were especially useful. My thanks are due also to Dr F. R. Burden, Monash University, for permission to use certain results obtained in his laboratory, to Mr A. L. H. Smith, Deputy Headmaster, Carey Baptist Grammar School, for helpful comments, and to the *Scientific American,* and the Australian Academy of Science for permission to quote from publications. I am especially grateful to Mr Robyn Williams of the Australian Broadcasting Commission Radio Science Unit for much help with the original broadcasts on which this book is based and to my former employers, Kraft Foods Limited, Melbourne, for the continuing use of the facilities of their technical library. Finally, may I express my warmest thanks to my friend Professor John Hawthorn, a past president of the International Union of Food Science and Technology, for so willingly providing a Foreword.

It remains only to say that the responsibility for this book and the views expressed in it are mine alone. They are in no way attributable to any of the organizations with which I am or have been associated, or to any of those individuals whose help I have gratefully acknowledged.

K. T. H. Farrer
Blackburn, Victoria
January 1987

A Note on Units and Concentrations

S.I. (Système International) units are used throughout this book thus:

I kilogram = 1000 gram (g)
= 1 000 000 milligram (mg)
= 1 000 000 000 microgram (μg, also written mcg)
= 1 000 000 000 000 nanogram (ng)
= 10^3 g = 10^6 mg = 10^9 μg = 10^{12} ng

The unit of volume is the litre, l, to which millilitre, microlitre and nanolitre bear the same arithmetical relationship as milligrams etc. do to *grams,* not kilograms.

The larger units, g to kg, are familiar through ordinary purchases and the cook is familiar with smaller quantities. Thus, a standard level teaspoonful of ordinary sugar weighs 4 g; of table salt, 5½ g. One centimetre cube of butter weighs about a gram and a half and enough sugar just to cover one new penny weighs only ⅓ g, but all these quantities, even the last, are large compared with that of a contaminant present in a food and which is written as a concentration.

Concentrations are expressed as weight of substance quoted per weight of reference substance or as volume per volume, or, sometimes, as weight per volume if that is convenient: for example, 1 mg of, say, a heavy metal per 1 kg of food or tissue. Such a concentration is, from the above table, 1 part per million (1 p.p.m.): 1 μg/kg is one part per thousand million, which the Americans call a billion, often expressing this concentration as 1 p.p.b. This is, however, misleading and modern practice is to eschew p.p.m. and p.p.b. and to use the above units about which there can be no ambiguity.

1 l of water weighs about 1 kg and, therefore, concentrations expressed as, say, milligrams per litre (mg/l) of water or sea-water or even such

biological fluids as milk and blood, may, within the limitations of sampling and analytical methods, be regarded as equivalent to milligrams per kilogram, i.e. to parts per million in the old terminology.

Concentrations stand separately from absolute quantities. Thus, if an ordinary brick veneer suburban house requires, say, 12 500 bricks, and a developer builds, say, eighty houses in an outer suburban building estate, then one can stand in the midst of them holding one brick which represents a concentration of one part per million. Alternatively, if one stands beside a home swimming pool, 10 metres by 5 metres with an overall average depth of 1½ metres, 75 millilitres (about 2½ oz) of water held in a glass in the hand is equivalent to 1 p.p.m. of the contents of the pool. Or, put another way, a substance to be added at a concentration of 1 p.p.m. (1 mg/kg) to keep the pool free from slime would require the addition of 75 g (or 2⅔ oz) for the whole pool.

There is no difficulty in visualizing or, indeed, measuring these quantities, but suppose there is a carry-over of mercury into a 100 g serving of fish at a concentration of 0.2 mg/kg, then the mercury in that serving is 20 μg. Such a quantity would barely be visible, nor can it be weighed easily, and trying to detect 1 μg/kg (1 p.p.b.) of aflatoxin in peanuts has been likened to looking for a needle in 256 one-ton haystacks. But such concentrations may be handled comfortably in the laboratory. This is done by weighing a gram, say, of a substance, dissolving it in a large measured volume of a liquid, usually water, then, by making suitable measured dilutions of this solution, taking a few micrograms or even nanograms of the particular substance by which the chemist is able to standardize the specialized equipment used to estimate very low concentrations of it as, for example, the 20 μg of mercury in the 100 g of fish or the even lower concentrations and quantities of aflatoxin in a consignment of peanuts.

1 cm³ butter
≈ 1½ g

enough sugar
to cover
one new penny ≈ ⅔ g

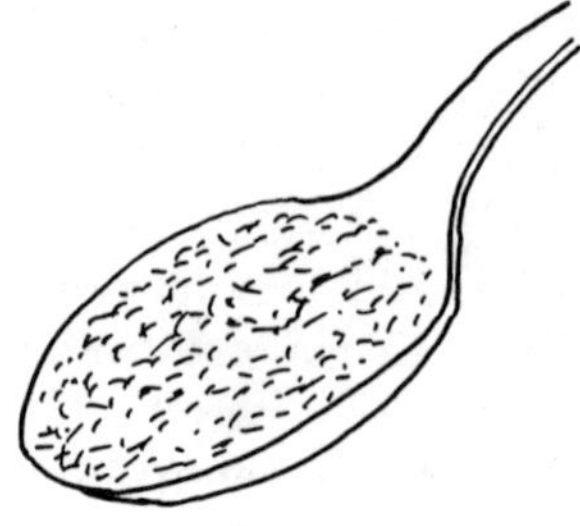

a standard level
teaspoonful of
salt ≈ 5½ g; of
sugar ≈ 4 g

Abbreviations

ABS: acrylonitrile-butadiene-styrene polymer
ADI: Acceptable Daily Intake
CAFTA: Council of Australian Food Technology Associations
CIP: cleaning-in-place
CSIRO: (Australian) Commonwealth Scientific and Industrial Research Organization
EEC: European Economic Community
FAC: Food Additives Committee of the NHMRC, Australian Commonwealth Department of Health
FAO: United Nations Food and Agriculture Organization
FSC: Food Standards Committee of the NHMRC, Australian Commonwealth Department of Health
FST: Food Science and Technology (Reference) Subcommittee of NHMRC, Australian Commonwealth Department of Health
GRAS: Generally Recognized as Safe (by the United States Food and Drug Administration)
IAEA: International Atomic Energy Agency
IOFI: International Organization of the Flavour Industry, Geneva
JECFA: Joint Expert Committee on Food Additives of FAO/WHO, Rome
JECFI: Joint Expert Committee on the Wholesomeness of Irradiated Food of FAO/IAEA/WHO
MSG: monosodium glutamate
NHMRC: National Health and Medical Research Council, Department of Health, Canberra
OECD: Organization for Economic Co-operation and Development
PBB: polybrominated biphenyl
PCB: polychlorinated biphenyl

PER: protein efficiency ratio
PTWI: Provisional Tolerable Weekly Intake (of an additive or contaminant)
PVC: polyvinyl chloride
VCM: vinyl chloride monomer, the monomer from which the polymer polyvinyl chloride is made
WHO: United Nations World Health Organization

1 Reasonable Doubt

Where doubt is, there truth is – 'tis her shadow.

P. J. Bailey, *Festus*, Book V

Whether the informed professional likes it or not, and he doesn't, the fact remains that today there is widespread doubt about food additives and contaminants. In recent years and in ways quite unjustified by the body of information available, the emotions of many people in the western world have been stirred. Food additives, especially, have been widely condemned, but food regulatory authorities throughout the world, when considering what manufacturers deliberately add to food, have been faced with the even more important and pressing question of what substances find their way into food by accident.

In the history of the food processing industry dangerous additives *have* been used. Dangerous contaminants *are* found in the modern food chain; but there is no justification whatever for the scaremongering books, newspaper articles and television programmes which, out of ignorance, or for gain or notoriety, periodically attack and denigrate the food supply of our society. Vigilance is justified and necessary; vituperation is not.

As the poet implies, doubt is often the first step towards truth. What is needed, therefore, in this debate is the cool assessment of scientific facts, not strident warnings about the 'chemicals in our food' which ignore the chemical nature of all food and body tissues and the thousands of continuing biochemical reactions without which life would be impossible. Foreign molecules may interrupt some of these reactions and, if in sufficient concentrations, lead to visible effects in the behaviour of the individual, i.e. to symptoms of poisoning; but for the most part they are dealt with quickly by detoxifying mechanisms, and eliminated from the body.

Very rarely do popular reports about food additives and contaminants talk about concentrations, either those in food or those necessary to produce the effects described. Since the 1950s dramatic advances in the techniques and instrumentation of chemical analyses have pushed the limits of detection of a great many chemical compounds to concentrations lower than was ever imagined. Detection and measurement of substances in micrograms per kilogram (parts per thousand million) or even micrograms per thousand kilograms (parts per million million) are now commonplace, with the result that compounds, the presence of which had never been suspected, are now being found in all kinds of foods. They have, in many cases, always been there, like mercury in sea-water and in free-ranging ocean fish. What they mean in terms of public health is often unclear but, having been found, they must be accounted for.

Beer, wine and spirits introduce a foreign substance, alcohol, into the body. Everyone knows that a drink before dinner and a glass or two of wine during the meal produce no visible effect in most people. Equally, everyone recognizes the early and later stages of 'in-toxic-action' (that is, poisoning; in this case by ethyl alcohol). In other words, the body is able to cope with small concentrations of the compound, but the swamping of the metabolic processes for dealing with it leaves it free to affect the brain and, if the warnings are not heeded, death will result. This is a familiar example of dose/response. A given dose will produce a given response and this, in essence, is the basis of toxicological studies.

The reports of such scientific studies always carefully describe the experimental conditions; and the interpretation of the results requires the skills and special knowledge of the toxicologist. Too often popular reports try to deal with information being obtained on the frontiers of knowledge and still subject to confirmation and refinement, and then to present it in an over-simplified way.

Further, there is a lack of understanding of the enormous advances in our knowledge of the chemical nature of foods and food constituents and of the curiosity which drives the chemist to duplicate in the laboratory what nature does in the plant or animal. Thus, vanilla and vitamin C, both natural products, have within living memory been synthesized in the laboratory and are now made in factories. They are therefore totally synthetic articles of commerce; but they are not artificial. They are no different from the natural substances extractable at greater cost from natural sources.

Similarly, the study of the interrelationships of the chemical and physical properties of foods has revealed the underlying scientific basis of food texture, food emulsions such as cake batter, and the complex phenomena met with in bread doughs. The unravelling of the chemistry of fats and oils, a twentieth-century achievement, has permitted the food technologist to specify very closely the properties he needs in cooking oils and in fats for baked products. The changes brought about in the molecules of, say, a vegetable oil by the addition of atoms of hydrogen, yield a product closely approximating the composition and texture of butter, and the addition of vitamins A and D and a little skim-milk powder gives margarine. But does anyone regard margarine as synthetic or artificial? Chemical knowledge has allowed the technologist to take a natural product, vegetable oil, and by simple chemical manipulation to change its properties so that it resembles another natural product both in texture and nutritional value. So also with other less familiar things.

Historically, food contamination preceded the use of food additives. Man, the hunter-gatherer, ingested natural poisons and contaminating micro-organisms beyond our experience. He also consumed charcoal, mineral matter and carcinogens (cancer producing substances) with his singed and charred meat.

By trial and error he developed special knowledge of the suitability of foods available to him and of ways of treating potentially dangerous berries and roots to make them edible. By trial and error he learned, too, how to dry food and to use salt to protect fish and fermenting animal and vegetable materials from the damaging growth of dangerous micro-organisms. In other words, a rough and ready food technology developed and, over a long period of time, gave man the means of avoiding most of the immediate consequences of eating contaminated food. As we shall see later, much contamination by slower acting and therefore unrecognized substances and organisms must have occurred, but for centuries empirically developed methods of food handling and treatment provided a measure of protection.

It would be a mistake to imagine that the use of food additives is a modern development. Chemical preservatives were probably the first food additives to be used and salt has been so used since ancient times. Wood smoke, the resins of resinated wine, and the pitch lining of barrels for wine and beer are examples from classical and medieval times of the unknowing use of chemical preservatives, but very early in his history man realized the importance of colour, flavour and texture. The Romans were thickening sauces with wheat starch. They also yearned after white bread and generously added powdered alum to get it, not as an adulterant but as an additive; and as late as the turn of this century bakers were being prosecuted for using it for the same reason.

The use of spices as flavourings is very old also, especially in the eastern lands in which they are grown. A spice trade developed with Europe in ancient times, was large by the first century A.D. and 3000 lbs

(1360 kg) of pepper was part of the price asked for raising the siege of Rome in 408 A.D. Spices were popular in England from Norman times, derived wisdom attributing their use to their value in masking the flavours of tainted and salted meats. True or not, the demand for strong, aromatic flavours was established.

In this period colours derived from plants were being used: indigo, alkanet from the root of the borage, sanders from sandalwood and saffron; black was obtained from blood. Cochineal, prepared from the insect, *Coccus cacti,* was in use in Europe at least by the early seventeenth century. A hundred years later the emphasis for food for the tables of the wealthy was on colour rather than flavour but, apart from cochineal, the origins were vegetable: yellows of saffron and marigold, greens of sage and spinach, pinks, blues, mauves and violets of petal extracts, and by the late eighteenth century annatto from the pods of the shrub, *Bixa orellana,* was introduced from the West Indies to colour cheese. Annatto is structurally related to vitamin A and is still sometimes used as a cheese colour.

Saffron was used as a flavour as well as a colour, so was turmeric; and cinnamon, ginger, pepper and others less familiar to modern tastes were the flavour additives of the time. They are now so well accepted as to be separately described in food formulation as spices or condiments.

These additives were used by the chefs who cooked for the great houses, but expectations had been aroused; for the agricultural revolution in the earlier part of the eighteenth century was no less important than the industrial revolution, and food in England improved in quality as well as quantity. More meat was available; the demand for white wheaten bread became general; and tea, which had been introduced to England in the mid-seventeenth century, became a general beverage throughout the whole country by 1740.

Until the eighteenth century there was virtually no food processing industry except, perhaps, the centralized grinding of corn at the manorial and, later, the village mill, so food additive use was confined to the kitchens – mainly, as already noted, those of the big houses and castles; but, in any case, food production was local, serving small communities and everyone knew what the miller, the baker and the brewer were up to. Later, the towns began to grow and it became necessary to develop some kind of food supply to feed the new factory workers.

The growth of towns, the wars with France culminating in the Napoleonic period, poor harvests and the lack of a food industry which could effectively link times of plenty with times of want brought Britain to the brink of famine in 1812 and made the whole of the first quarter of the nineteenth century a period of want. Food processing became impersonal as country producers supplied distant townsmen via middlemen, agents and processors; and as the nineteenth century progressed, the processors became larger and food and drink began to be prepared in factories for sale to people who could no longer grow and prepare their own.

The earliest food manufacturing processes were brewing and baking, the former being the first to be centralized. In the eighteenth century beer was being coloured with tobacco, treacle and molasses and flavoured with all kinds of things, including the poisonous Cocculus India Berry (the seeds of *Cocculus indicus,* a poisonous drug used medicinally). At the same time, bread was being whitened with alum, and, as a foretaste of things to come, or perhaps, simply as a commentary on unchanging human nature, there was a rumbling accusation, which lasted for half a century, that bakers were also using chalk, whitening and ground-up bones from the charnel house. There was no doubt about the alum; it was still being used at the turn of this century, but it was pointed out in 1758 by H. Jackson (see below) that the other three were easy to detect, reduced the loaf size and damaged the crumb texture.

The adulteration of food is as old as the selling of it and the increasing volume of food coming out of factories and shops had provided more opportunity for deceit.

> Food adulteration [said Burnett] was peculiar to an age of rapid industrialization and urbanization . . . By the closing decades of the eighteenth century, however, the quality of many foods was rapidly deteriorating, and it is beyond dispute that for most of the next hundred years adulteration was an exceedingly widespread and highly remunerative commercial fraud.

While much of it was aimed simply at dilution with a cheaper substitute some involved what we would today call additives – colours, flavourings and thickeners. There was virtually no knowledge of the chemistry of foods and consequently of the analytical methods for its examination. There was therefore no control either of food or of the use of food additives.

The first attempt to describe food adulteration by chemical analysis dates back to 1758 when Jackson, describing himself as a chemist, published a little book called *An Essay on Bread; wherein the Bakers and Millers are vindicated from the Aspersions contained in Two Pamphlets.* Strangely enough, its main thrust was the defence of bakers against charges of gross adulteration.

The first revelation of irregularities was Accum's *Treatise on the Adulteration of Foods and Culinary Poisons* published in 1820. While he was concerned primarily with the adulteration of foods, his examinations showed that additives were being used to deceive. The ancient use of alum to whiten bread was commonplace, copper salts were being used to colour pickles and red lead to colour Gloucester cheese. Brewers were using deleterious substances; even, as already noted, *Cocculus indicus,* falsely to simulate strong beer, and so on. Accum was hounded out of Britain and the furore caused by his revelations died down.

In 1848 John Mitchell published his book *A Treatise on the Falsifications of Food, and the Chemical Means Employed to Detect Them.* Like Accum, he was concerned primarily with the adulteration of food, and there was a lot of it, which is outside the immediate scope of this book, but his analyses showed the presence of many deleterious food colours; black lead, Prussian Blue, lead chromate, copper carbonate and, of course, alum to whiten bread. This time something happened. The *Lancet,* the English medical journal, established an Analytical Sanitary Commission and in 1855 it confirmed considerable adulteration of food; alum in bread, naturally, and red lead, Prussian Blue, Brunswick Green, vermilion and copper arsenite as colours in confectionery. There was a good deal more and legislative action began in Britain in 1860.

Things were no better in Australia and in 1863 the Melbourne *Argus* listed many common adulterations. Red paint to colour anchovies and quassia to make beer bitter were two of them – additives used to deceive. Worse, however, was the revelation in 1876, again by the *Argus,* of analytical results which showed that of 69 samples of orange or yellow sweets on sale in the Melbourne area 64 were coloured with lead chromate. The authorities lacked the legislative and policing power to deal with this scandalous situation, but the *Argus* stopped it by public exposure and by follow-up publicity a year or two later.

At about the same time Melbourne newspapers exposed the practice of collecting spent tea leaves and 'facing' them with a mixture of Prussian Blue and turmeric so that they could be resold as green tea. Legislative action in Victoria in 1881 put a stop to that in the Australian trade and in 1883 new legislation provided for the proper inspection of the food supply.

Attention then turned to preservatives. Boric (boracic) and salicylic acids were the favourites at that time and were used in such commodities as sausages, milk, butter and cordials. Earlier, however, from the 1830s to 1870, many attempts had been made to use sulphites, particularly phosphoric and lactic acids, acetates, nitrates and nitrites to preserve meat for periods long enough to ship it to Britain. Other attempts, bizarre in modern eyes, included a system for causing animals to inhale carbon dioxide before slaughter and another in which a pint of blood was drawn from the animal, mixed with boric acid solution and 'transfused' back into the beast through the jugular vein. After two minutes to allow for distribution throughout the system, the animal was slaughtered.

Needless to say, these proposed uses for additives were doomed to failure and attention at the end of the century was concentrated on the use of preservatives in what were, given the current state of knowledge and technology, legitimate ways, but for which there was no provision under the legislation forbidding the adulteration of food. Indeed, the situation became so tense and confused that one magistrate said, 'it would be well if the question of preservatives was settled by scientists one way or the other so as to protect manufacturers from annoying prosecutions'.

In the United States of America, Dr H. W. Wiley, chief food chemist of the U.S. Department of Agriculture, attempted to settle the question with his 'Poison Squad' of volunteers, and gained valuable information about a number of common preservatives. In Victoria the Government Analyst, Percy Wilkinson, had begun in 1899 to gather analytical information about food adulteration and much of it related to the use of food additives; colours, preservatives, sweeteners (saccharin), flavours – and alum. The *Argus* was already differentiating between food additives used to deceive, such as yellow colours to simulate egg in baked goods, a practice dating from the Middle Ages, and those that were harmful, but Wilkinson's report was the final stimulus required and at the end of 1905 Victoria enacted the *Pure Food Act* and led the world into the establishment of food regulations. The rest of the Australian states followed quickly and right from the beginning the philosophy was exclusive, that is to say, those food colours and preservatives which could be used were written down in the regulations; all others were excluded. This was in contrast with the practice which developed in the United Kingdom in which ostensibly anything could be used except those substances which were specifically forbidden; and in the United States there grew up the GRAS list of substances which were 'generally recognized as safe' on no other grounds than that they had not been shown to cause any damage to human beings. From this list the manufacturer could use any substance in any food. The Australian and New Zealand approach from the beginning was philosophically the antithesis of these attitudes; it was ex-clusive rather than in-clusive. Today, the Americans are far more demanding of GRAS status and the United Kingdom has adopted permitted lists for additives.

The concern over preservatives at the turn of the century has already been referred to, but at that time some eighty textile dyes, products of the dye-stuffs industry which dated from Sir William Perkin's accidental discovery of mauveine in 1856, were being used in foods. The concentrations were small, and, while no one thought to question any long-term effects, it was obvious that they were immeasurably safer than the lead and other metal salts so recently used in confectionery. Regulations introduced in various countries in the early part of this century reduced the number of dyes being used in food, but no one thought much about additives until after World War II. Concern over additives and contaminants dates from the 1950s and coincides with dramatic advances in analytical chemistry, greatly increasing knowledge about cancer and its environmental associations, and the establishment of international agencies to consider food and the regulation of it in world trade.

At first, food additives gained the most attention because they were identified as substances being deliberately added by manufacturers to food, and even some government people who should have known better assumed that manufacturers were *ipso facto* irresponsible and their technical staffs lacking in scientific probity. Happily, all round the world the regulators have long since realized that the food processing industry *is* responsible, its staff frequently better informed scientifically than they are and that in its own interests it will go to any lengths to avoid the association of its products with any kind of sickness.

With additives, the matter is firmly in the hands of the manufacturer. He adds or he does not add. Contaminants are another matter. They come uninvited and unseen. Many of them have always been with us and some of them are part of the very food itself, not so much contaminants as inbuilt hazards. We begin with these, pass on to the contaminants, and then consider additives and the rationale of the assessment of the risks they pose. By far the greatest public health hazards come from microbial contamination and these are considered before the final discussion of the control in the food supply of all additives and contaminants.

2 Toxicants Naturally Present

Nothing to do but work
Nothing to eat but food
Nothing to wear but clothes
To keep one from going nude.

So wrote 'Bow Hackley' (Benjamin Franklin King), an obscure American humourist of the late nineteenth century. 'Nothing to eat but food', as, for example, the ten thousand Greeks fighting their way back home from Babylon under Xenophon in 401 B.C. When they came to a Greek colony on the Black Sea coast, 'for the most part', wrote Xenophon in the *Anabasis,*

> there was nothing here which they really found strange, but the swarms of bees in the neighbourhood were numerous, and the soldiers who ate of the honey all went off their heads and suffered from vomiting and diarrhoea and not one of them could stand up –. So they lay there in great numbers as though the army had suffered a defeat and great despondency prevailed. On the next day, however, no one had died, and at approximately the same hour as they had eaten the honey they began to come to their senses; and on the third or fourth day they got up as if from a drugging.

'Nothing to eat but food', but what had happened? There was an abundance of rhododendrons in the neighbourhood and the bees had concentrated in the honey the poisonous andromedotoxins of the flowers. Other examples of poisonous honeys are known today. In New Zealand honey gathered at certain times from the tutu is acutely poisonous and harvesting is controlled by regulation. In Australia honeys from *Echium plantagineum L.* (Patterson's Curse or Salvation Jane) have been shown to contain traces of pyrrolizidine alkaloids which are known carcinogens. The public health significance of this is as yet unknown; more work is required.

As the bees passed on to man the toxins gathered from plants, so may cows pass on toxic substances in milk. Formerly, for example, cows eating the North American snake weed secreted in their milk a toxic alcohol, tremetol, which caused deaths in humans. When the source was identified in 1917, the problem became a minor one because the weed is easily recognized and readily removed from the pastures.

Such cautionary tales make the Promised Land 'flowing with milk and honey' sound less attractive. Fortunately, however, the poisoning of such wholesome and desirable foods by the concentration in them of unpleasant substances from the raw materials used by their respective biological factories is most uncommon. On the other hand, they do remind us that substances dangerous to man may appear in food independently of human activity.

We shall look at some of these compounds in the food groups with which they are associated, but, first, what do we mean by toxicant and toxic, poison and poisonous? The two sets of words are interchangeable; a toxicant is a poison, toxic is poisonous. A toxin, on the other hand, is a poison produced by a living organism, particularly a micro-organism. Thus, arsenic is a poison but not a toxin, but the extremely dangerous physiologically active metabolite of *Clostridium botulinum* is both. All these words have the connotation of rapid action but in Chapter 1 we referred briefly to the body's ability to deal with foreign molecules. We noted that poisoning follows when the detoxifying mechanisms available are swamped and, therefore, that the *concentration* of a substance in the food eaten is what determines whether or not, at a particular time, it will be poisonous; that is whether or not it will kill or injure health. We shall examine this in more detail in Chapter 5.

In this chapter we shall look at some foods which may be poisonous in the accepted sense and at others which contain small concentrations of substances normally considered to be poisonous but which are themselves harmless when used as part of a normal varied diet. We shall also discuss some naturally occurring carcinogens and other substances which have a long-term action, but we shall reserve for a later chapter consideration of food poisoning micro-organisms and the mycotoxins produced by various moulds.

Fish and Shellfish Toxins

Fish, too, may become toxic by virtue of their diet. Ciguatera poisoning from many different kinds of fish, mainly in tropical waters, is now thought to be due to an organism, *Gambierdiscus toxicus*, which produces in the fish a toxin which causes the flesh of the fish to become poisonous. The problem is a serious one for the populations of a number of Pacific islands. Fortunately, a test for the toxin has recently been suggested. A dinoflagellate species, which produces the so-called 'red tides', leads to a concentration in shellfish of saxitoxin, the cause of 'paralytic shellfish poisoning'. In both cases it is the animal itself, fish or shellfish, which has become poisonous to man. Altogether, about a hundred families of fish have been reported to be poisonous to a greater or lesser extent. Many of them, as already stated, become poisonous from what they eat. Others produce toxins naturally either consistently or at certain times of the breeding cycle. Others again have poisonous skins which must be removed before cooking.

Some fish, if improperly stored, are susceptible to a bacterial decomposition which leads on ingestion of the fish to symptoms similar to a severe histamine reaction, that is, to a severe attack of asthma. This condition is called scombroid poisoning because it is frequently associated with tuna, mackerel, sardines and other members of the family *Scombridae*. It is thought to be due to the conversion by the bacteria of the amino acid histidine into histamine well before there is any sensory evidence of spoilage. Scombroid poisoning is quite different from the so-called food poisoning (see Chapter 6) but must, nevertheless, be very close to the borderline between natural toxicants and toxins produced by contaminants.

One of the most notorious fish stories is that of fugu, a Japanese delicacy prepared from a species of the blow or puffer fishes, only parts of which are toxic. Puffer fishes are found all over the world and the toxins are located in the gonads and associated viscera particularly during the spring and early summer. The Chinese and Japanese have eaten these fish for thousands of years and many have died through ignorance or misadventure. Few other people will touch puffer fish which are unattractive, anyway, but the Japanese, carefully dissecting gonads and liver, prepare the flesh in specially licensed restaurants and serve it raw and paper thin. Though rare today, fatalities caused by carelessness in preparation do occur and the element of risk in eating fugu may be one of the reasons why the dish is so popular with Japanese gourmets.

Poisons from Plants

The vegetable kingdom provides the greatest variety of natural toxicants in food, some acute, some chronic; some must be removed to make a product edible and some are present in concentrations which may be ignored in a mixed diet. Hydrocyanic or prussic acid (chemical formula, HCN) acts very rapidly and with its associated odour of bitter almonds was a favourite with writers of detective stories. It was used by a number of Nazi leaders to avoid capture. Linked with sugars as glycosides, it occurs naturally in a number of vegetable products used for human food. For example, the bitter varieties of cassava or manioc (*Manihot utilissima* and other species) may contain up to 1500mg/kg (0.15%) of HCN equivalent in the dry solids and when the cells are broken by chewing, say, or grating, an enzyme liberates the cyanide as free HCN. Such glycosides are called cyanogens.

In the western world cassava is known almost solely as a convenient substrate for the production of alcohol and other fermentation products, but in developing countries it is the staple food of some 300 million people. It is particularly important in West Africa where millions of tonnes are grown annually. Though poor in protein it is rich in starch and supplies very high quantities of energy per unit area of cultivation. For generations cassava was rendered edible by thorough leaching with water and other traditional methods such as the preparation of gari in Nigeria. Gari is a grit which is made by grating, removal of excess water, fermentation of the grated pulp and roasting of the final mash. Such a process washes out much of the cyanide and the roasting drives off more. A great deal of research into the detoxification of cassava is going on in the areas where the plant is important as a food and recent work in Nigeria on bitter (1375mg of HCN/kg dry weight (d.w.)) and sweet (170 mg/kg d.w.) cassava has shown that in preparing gari from both varieties over 98 per cent of the HCN can be removed and, by using a 24-hour screw pressing, reduced in each case to less than 5mg/kg d.w. compared with about 12 by the traditional method. Similar work in Sri Lanka has given similar results.

Cassava is not part of the western food pattern. Of more concern is the amygdalin content of almonds and the kernels of apricot, peach, nectarine and other edible nuts. This glycoside is hydrolyzed by an enzyme, emulsin, to glucose, HCN and benzaldehyde, and there was for a long time in Australian food regulations an upper limit of 50 mg/kg (parts per million) of HCN in marzipan and almond paste and in the imitations of these two products made from other raw materials, especially apricot kernels. Recent work has suggested strongly that the regular consumption of low levels of HCN can produce neurological disorders and attention was therefore directed to apricot kernels which are freely offered for sale in so-called health food shops and which form the basis of the alleged cancer cure, laetrile, promoted so strongly in the United States. Accordingly, in Australia the permitted upper limit for HCN in marzipan and related products has been reduced to 5mg/kg and attention drawn to the danger of the regular consumption of apricot kernels. It had been thought that if the emulsin which frees the HCN could be destroyed by heating then the apricot kernels could be eaten with impunity. It now seems, however, that there is in the human alimentary tract another enzyme which can release HCN, so the original concern about this product remains.

Lima beans contain 0.01 to 0.3 per cent of HCN so that a meal made entirely from them could be hazardous. In a mixed diet there is no danger as the concentration of HCN is diluted to safe levels.

Other legumes contain undesirable compounds which are potentially dangerous. Thus, broad beans contain a nucleoside, vicine, and if eaten raw may induce an acute haemolytic anaemia, called favism, in individuals with a certain inherited metabolic defect. Favism is quite common in the Mediterranean area.

Also well known in that area and common in India and some parts of Africa is lathyrism. A degeneration of the spinal cord leading to muscular weakness and paralysis of the lower limbs, it is in two forms; neurological, which affects the central nervous system, and, in animals, osteological, which affects the bone and connective tissue. It follows a high level of consumption of the chick pea, *Lathyrus sativus,* and is due to an oxalyl-substituted amino acid in the seeds. This pulse is frequently available when other foods are not and the high intake which leads to lathyrism is often the only alternative to starvation.

In the seeds of many legumes there is a group of plant proteins which can cause red blood cells to clump, or 'agglutinate'. Because of this, they are called haemagglutinins, and ricin, from the seeds of the castor bean, is probably the most toxic of them. The toxicity of these beans and of castor oil press cake has been known since 1889 but most of our knowledge of the properties of haemagglutinins from many legumes, potatoes and cereals is less than thirty years old and details of their toxicology in man is sketchy. Fortunately, plant haemagglutinins are deactivated during cooking, but where raw legumes are used, or cooking of a staple food is incomplete, toxic effects such as diarrhoea and poor growth in children have been observed. Favism is a special case of haemagglutinin toxicity in susceptible individuals. Soybeans contain haemagglutinins and other toxins, including goitrogens, as well as allergens, yet the soybean is probably the most important oilseed on the planet and, as bean curd, is a source of protein and calcium for many millions of Asians.

Lupins are very old food plants, the seeds of the white variety having been used for at least three thousand years as a source of protein in the Mediterranean region from the Iberian peninsula to the Levant. Another variety has similarly been used in the Andes of South America. Of all the vegetable crops lupins supply the highest percentage of protein but they contain significant concentrations (1.5%) of poisonous alkaloids which can lead to depression, muscular paralysis and death from asphyxia, and lupins used for fodder have been shown to be teratogenic, i.e. to produce serious birth defects, in calves. The seeds, therefore, were suitable for use as food only after thorough leaching in water, up to days at a time, to a residual total alkaloid concentration of, say, 0.1 per cent.

Because lupin seeds are such a useful source of protein, breeding programmes, one of the most successful being in Western Australia, have resulted in the establishment of varieties in which the natural alkaloid content of the seeds has been reduced to as low as 0.1 per cent. The associated bitterness has gone and these 'sweet' lupins are potentially valuable sources of human food. In Chile, for example, the protein efficiency ratio (PER) of lupin flour, supplemented with the sulphur-containing amino acid, D,L-methionine, has been shown to be comparable with that of casein, the protein of milk, and the introduction of lupin flour into substituted infant milk formulae and other foods is being explored. Suggestions for similar applications in Australia have been made.

Potatoes, also, contain an alkaloid, solanidine, which is linked with three sugars to form a toxic glycoside, solanine. It occurs at concentrations of 0.01-0.10 per cent of the dry weight of the tubers but twice as much may be found in the peel. Solanine increases up to seven times when the tubers are allowed to turn green in light and it concentrates in the eyes and shoots. Levels greater than 20mg per 100g of potato (0.1% of the dry weight) are considered to render potatoes unfit for human consumption. Solanine is neither leached out nor destroyed during cooking. There are cases on record of potato poisoning but peeling removes most of the solanine and avoidance of eyes and shoots is prudent. Most of the reports of trouble relate to the use of potato sprouts as edible shoots.

As already mentioned, pyrrolizidine alkaloids have been identified in some Australian honey gathered from *Echium plantagineum L.* More common sources of these alkaloids are species of *Senecio, Heliotropum* and *Crotolaria* and herb teas made from them and given to children as medicine have caused serious illness and even death. Cases have been reported from Jamaica, India and the United States. The possibility of the transmission of these unpleasant compounds to humans via the edible products from animals grazing on the plant sources has not been overlooked but evidence, one way or the other, is lacking.

Mushrooms and other edible fungi have been eaten for thousands of years. A few of the many species are poisonous, some of them lethal, and the toxins belong to several different groups of chemical compounds. The toxic species are generally well known but most people in any case confine their attentions to the common field mushroom, *Agaricus campestris,* and, especially, to the cultivated variety, *A. bisporus*. The most common mushroom 'poisoning' is an allergic reaction, sometimes even to the cultivated variety.

Oxalic acid is a strong acid, corrosive and poisonous. It is found in plants as the sodium, potassium and calcium salts. Rhubarb and spinach may contain up to 1 per cent and even more of oxalates on the fresh weight and tea leaves up to 2 per cent, but no food regulatory authority anywhere would permit the addition of oxalates at any level to a formulated food.

There are records of sickness and death following the abnormal ingestion of both cooked and raw rhubarb leaves and stalks, the cause being attributed to the oxalic acid normally present. The evidence is not conclusive, however, and there is grave doubt about this attribution because it seems at least to be likely that toxic anthroquinone derivatives also known to be present are involved. At the same time, it cannot be gainsaid that some plants used for food contain significant concentrations of an acid known to be harmful to man and animals.

Calcium oxalate is excreted in the urine of normal healthy people but under abnormal conditions may precipitate and form kidney stones. It comes from a number of metabolic sources including vitamin C and there is at least presumptive evidence that regular taking of massive doses of this vitamin, particularly with a high intake of calcium, can substantially increase urinary oxalate and favour the build-up of kidney stones.

Nitrogen is a keystone of plant growth and the inorganic form occurs in plants as nitrates, particularly, and nitrites. The presence of these compounds is influenced by many factors including the variety of plant, the amount of nitrate available to it either naturally or as fertilizer, drought and so on. Nitrate poisoning was recognized first in cattle in 1895 and was attributed to extraordinarily high levels of potassium nitrate in fodder plants growing on highly manured soil. The nitrate was reduced to nitrite in the rumen and methaemoglobinaemia resulted.

Certain vegetables used for human food tend to contain higher concentrations of nitrate than others. Spinach and lettuce are examples but there is great variation. The acute poisoning of infants by nitrates in spinach was reported in 1964 and other cases have been documented since. It is attributed to the bacterial formation of nitrite in raw spinach high in nitrate. Shredded spinach stored for one day at 25° C has been found to contain 1500 mg/kg of nitrite but freezing inhibits this reaction and heat processing stops it. Nitrite poisoning has been demonstrated in others who seem particularly sensitive to it but the degree of sensitivity and the distribution of it are not known.

It is known, however, that nitrites may react with secondary amines to form nitrosamines which have been shown to cause cancer in some laboratory animals. The direct application of such a finding to man is clouded by the undoubted facts that nitrites circulate normally in human saliva, that secondary amines are normally present in abundance in foods and that there is no evidence that nitrosamines act as carcinogens in humans either at all or at the low concentrations likely to be met with. The significance of the use of nitrates and nitrites as food additives, as distinct from normal constituents of foods, will be discussed in Chapter 4.

In contrast to all these examples, the akee fruit, a staple food of the poor in Jamaica, contains a toxic substance only when unripe. This compound is hypoglycin which produces severe hypoglycaemia but the hypoglycin disappears as the fruit ripens and the mature fruit is harmless. Formerly the mortality from eating the unripe fruit was 80 per cent but, once the cause was recognized, prompt injection intravenously of glucose ensured recovery. A similar phenomenon has been reported in grapefruit. A glycoside, naringin, irritating to the gastrointestinal tract, occurs in the green fruit but disappears as it ripens.

Goitrogens

The enlargement of the thyroid gland to produce the condition known as goitre is well known to be associated with iodine deficiency. Several dietary factors have been shown to contribute to this condition and a number of unsuspected components tested under the rather artificial conditions of the laboratory have been found to interfere with iodine absorption. The more important dietary factors are, however, a group of substances associated particularly with the brassicas (the cabbage family). The most potent of these has the rather forbidding name, 1-5-vinyl-2-thio-oxazolidone, and occurs in the seeds of many brassicas and, in a bound form, in the edible portions of turnip (*Brassica campestris*) and swedes (*Brassica rutabaga*). Cooking destroys the enzyme which releases the substance but cattle feeding on these brassicas transmit the activity to the milk. However, scalding and freezing reduce it and ice-cream, therefore, has less goitrogenic activity than the milk from which it is made. In areas of iodine deficiency such as Tasmania, where brassicas were used to supplement cattle fodder, the incidence of goitre was high and iodine supplementation of the diet was practised. However, iodine in excess is itself goitrogenic and when milk supplies were contaminated with iodine compounds (see Chapter 3) thyrotoxicosis was detected and iodine supplementation of the diet discontinued.

Carcinogens

In recent years a number of aspects of diet have been strongly correlated with cancer. Most of these associations are with contaminants of various kinds but carcinogens are known to occur naturally in some articles of diet.

Cycad nuts have long provided food for native populations in tropical regions but only after well known toxic constituents have been removed by repeated soaking of the 'meat' in water. As well as being acutely toxic, the untreated nut was shown to be very carcinogenic in rats probably because of the formation by bacteria in the gut of a carcinogen from a precursor in the nut extract.

Bracken shoots are used in salads or as greens in New Zealand, the United States and especially Japan; and recent epidemiological studies in Japan have shown a clear-cut positive relationship between intake of bracken fern and cancer of the oesophagus, thus offering some confirmation of the presence of a long suspected carcinogen. Interestingly, the same study suggests that a daily intake of green-yellow vegetables has a protective effect against several different forms of cancer.

Safrole, which is the major component of sassafras oil and a minor one of a number of other spice oils, is a weak carcinogen. Formerly used as a flavouring in root beer in the United States, sassafras oil has now been removed from lists of permitted additives in most countries.

Finally, studies in Australia are increasingly associating beverage alcohol with cancers of mouth, oesophagus and larynx and there is a presumptive correlation between beer consumption and cancer of the rectum.

Amines

Physiologically active amines such as tyramine, histamine, phenethylamine and serotonin are normal constituents of, say, cheese, chocolate, sauerkraut, wine, aged and fermented meats, bananas and so on. Thus, tyramine occurs in most varieties of cheese from low, say less than 50μg/g in Gouda, Provolone and processed cheese, to up to 2000 in some samples of Camembert and Stilton; in yeast extract which may exceed 2000; and in pickled herrings which may exceed 3000. Such concentrations injected would be disastrous, but when eaten they are harmless to most of us. The consumption of massive quantities of a particular product may, however, cause problems. Thus in West Africa a diet based on plantains contributes anything from 100 to 200 milligrams per day of serotonin, a level to which has been attributed the appearance of certain heart lesions. At concentrations encountered in a varied diet, tyramine could cause problems for susceptible individuals and histamine has been connected with food poisoning in one or two isolated and unusual circumstances. Both tyramine and histamine do, however, pose a threat to individuals being treated for depression with monoamine oxidase inhibitors. The metabolic pathway for dealing with the amines is blocked by the medication and blood pressure goes up.

There are reports, too, connecting tyramine, serotonin and phenethylamine with migraine attacks but much work remains to be done on this postulated relationship.

Fatty Acids

In recent years controversy over the contribution of diet to heart disease has focused attention on fatty acids and so much has been said and written that the terms saturated and unsaturated are often bandied about without any understanding of what they mean. Fats and oils are mixtures of triglycerides, i.e. compounds of glycerol with fatty acids which consist of chains of carbon atoms of up to more than twenty to which hydrogen atoms are attached and which terminate in the carboxylic acid group which gives them their name and which forms the link with glycerol. Where all the hydrogen atoms possible are attached, the carbon chain is said to be *saturated,* but where it is possible to add, say, two more hydrogen atoms, there is a 'double bond' in the chain which is then said to be *unsaturated;* in this case, *mono*unsaturated because there is *one* double bond. Where there are two or more double bonds, the chain is *poly*unsaturated.

This is not the appropriate place to discuss the several factors contributing to heart disease. Diet is only one of them, and not the most important one, and it is probably best to say only that current thinking is swinging away from the former emphasis on polyunsaturated fats towards a reduction of total energy intake and a reduction of the proportion of energy derived from fat – any fat. There is, however, one component of a natural oil which, for other reasons, must be considered.

Two important edible oils, from rape seed and mustard seed, contain a twenty-two carbon atom monounsaturated fatty acid (a $C_{22:1}$ acid) called erucic acid, at concentrations of from 20 to 50 per cent. Both oils are used extensively in some countries as cooking oils and in shortenings and margarine. From about 1960 work in several countries showed that diets high in rape seed oil produce changes in heart muscle in rats and other experimental animals and that the cause is probably erucic acid.

In a number of animals, including poultry and lambs, erucic acid from rape seed oil in the diet is transferred to the body fat, but it is unlikely to survive the rumen of cattle to appear in the milk. Although there is no evidence of harmful effects in humans, special breeding programmes in Canada to produce rape seed very low in erucic acid have been successful and rape seed yielding an oil containing less than 4 per cent has been grown in Australia since the mid-1970s. A regulatory limit of not more than 5 per cent in the oil has been set.

The only significant food use for rape seed oil is in margarine manufacture. More recently, the use in this industry of fish oils also has been proposed and has drawn attention to the presence in these oils of $C_{22:1}$ fatty acids, mainly cetoleic and erucic, at concentrations of 8 to 13 per cent and to the possibility that the effects in animals attributed to erucic acid may be common to all $C_{22:1}$ acids (called docosenoic acids) which accumulate in heart muscle and are only slowly metabolized.

Compared with some substances already discussed, the problem posed by erucic acid is almost esoteric, but it is clearly a case of a naturally occurring substance which *may* have caused problems in the past, which *does* harm some animals if eaten in high concentrations and which, as a precautionary measure, is better reduced to a minimum in human food in the future.

Caffeine and Related Alkaloids

Caffeine is a central nervous system stimulant with which tea and coffee drinkers have daily contact. It occurs in a number of plants which have been used from ancient times to make beverages. Man has thus been exposed to it and to its chemical cousins, theophylline and theobromine, for many generations. Tea, coffee, cocoa, cola drinks, the maté of South America, the guaraná and yoco of Brazil, and the infusions of the cassina of North America all contain caffeine. Tea contains theophylline and theobromine also and cocoa about three times as much theobromine as caffeine. Both caffeine and theophylline are potent stimulants, the concentrations of which in tea and coffee depend first on the concentration in the original tea or coffee and then largely on the strength of the beverages prepared from them. Caffeine is quoted at anything from 50 to 150mg per cup (a vague quantity) of tea and 85 to 150 per cup of coffee, even as much as 330 for well percolated coffee. Cocoa contains, say, 30 to 50mg per cup plus 95 to 150mg of theobromine.

Cola drinks, flavoured with infusions of the cola nut, also contain caffeine, say, 35 to 55mg per bottle or can, but it is said that at least half of it is added as the pure compound and the ethics of thus increasing levels of such a stimulant in a drink which is to be consumed very freely by children are at least questionable.

Recent studies in America have shown that caffeine can cause birth defects in rats. There is no evidence of any such effect in man but caffeine, theophylline and theobromine move freely about the body. The first two are known to cross the placental barrier to reach the unborn child and theobromine from chocolate rapidly passes into the mother's milk. It therefore seems prudent as the official American news release suggested for the expectant mother to 'put caffeine on her list of unnecessary substances', i.e. to cut down on coffee and tea and drink them weak.

Man's long contact with caffeine suggests strongly that the body can normally cope quite easily with moderate quantities but regulatory authorities are keeping a watchful eye on the current scientific discussion.

Vitamins and Anti-vitamins

Vitamins are trace components of the diet essential for the prevention of a number of diseases but the toxic effects of massive doses are well recognized. This is especially so of the fat-soluble vitamins. The polar bear, at the top of a food chain based on fish, stores massive amounts of vitamin A in its liver and cases of poisoning following a meal of polar bear liver are on record. And not only polar bear liver. During the 1911-13 Australasian Antarctic Expedition, Mawson, Ninnis and Mertz set out from Commonwealth Bay on a journey of exploration which turned into a nightmare. Six hundred miles from base, Ninnis was lost in a crevasse with the tent, all the dog food, most of the equipment and all but one week's supply of man food. Mawson and Mertz turned back and soon were forced to kill the dogs and eat their livers. Mertz perished and Mawson survived by the narrowest of margins. The dogs had lived on fish and it now seems certain that the men were poisoned by vitamin A stored at high concentrations in the dogs' livers. Normally the only danger of hypervitaminosis A is through the over-use of vitamin capsules, a practice which is causing increasing concern.

The other side of that coin is the presence in foods of anti-vitamins, antagonists of various kinds which interfere with vitamin absorption or which actively destroy vitamins. Of these probably the best known are the thiaminases which destroy vitamin B_1. These are found in raw fish and some plants, notably ferns such as the bracken fern consumed in Japan and elsewhere. Fortunately, anti-vitamins have little relevance in normal varied diets.

Food Allergens

There is another group of compounds which may scarcely be classed as toxicants yet which certainly have the capacity to injure the health of susceptible individuals. These are food allergens. Usually proteins, they may often be inactivated by heat; say, 30 minutes at 120 °C. Most foods have at various times been associated with allergic reactions but the most common are, probably, eggs, seafood, strawberries, milk, oranges and so on. Allergens affect only sensitive individuals and the mechanisms involved are yet to be explained; but allergens differ from poisons in another important regard in that the severity of the reaction depends not on the dose but on the degree of sensitivity of the individual. Identification of the cause of a reaction is often difficult and subsequent desensitization is frequently less than successful. The only sure way of avoiding allergic reactions is removal of the responsible food from the diet.

Another aspect of the allergy problem is the development in recent years of what is being called 'clinical ecology', or the recognition and treatment of distressing conditions said to be caused by the exposure of susceptible individuals to extremely small concentrations of substances of hydrocarbon origin. These substances are not naturally occurring toxicants, but are essentially contaminants, and some are said to be associated with additives. This will be discussed further in Chapter 5.

Other Plant Toxins

An account of this nature cannot be comprehensive and there are many other examples of toxic substances in foodstuffs. Gossypol, an acutely toxic yellow phenol from the glands of cottonseed, is a representative of a class, the plant phenols, not so far mentioned. It is of significance because cottonseed meal, made from the cake remaining after the oil has been expressed, is used as an animal feed and a raw material for the manufacture of certain human food supplements. Because of this, selective breeding has been used to produce cottonseed virtually free from gossypol.

One could go on to talk about the sulphur compounds called glucosinolates found in foods from the genus *Allium* (onions, leeks, chives, garlic and shallots) and which could be harmful if consumed in large amounts, or to point out that all foods contain trace metals, any one of which could be poisonous in high enough concentration. But the point is sufficiently well made that natural does not of itself, as some would have us believe, necessarily mean good or even harmless. Furthermore, it is clear that processing, of soybean and cottonseed, for example, as well as of other products already referred to, can turn hazardous raw materials into good food.

A Classification of Natural Toxicants

Table 2.1 is an attempt to develop a broad classification of naturally occurring toxicants in foods. In accordance with the desire to confine this chapter to toxic substances which are inherently part of the food and not the result of the contamination of it by micro-organisms or chemicals, all bacterial toxins, including mycotoxins, are excluded, as are heavy metal contaminations such as mercury, lead and cadmium. It may be argued that the inclusion of a toxin in honey, or even ciguatera poisoning in fish, is the result of contamination and that the methyl mercury in the tissues of free-ranging ocean fish from uncontaminated waters is a naturally occurring toxicant. So be it. Type I may be omitted if so desired, or retained and the mycotoxins added to it, but the metals may easily be accommodated in the classification under Type III (i) (b).

From this discussion the following are evident:

1. Some natural products were well known by tradition to be acutely harmful and were avoided while others were made safe by washing.

Table 2.1: A Classification of Naturally Occurring Toxicants in Foods

	Type	*Examples*	*Made safe by*
I	Toxicants concentrated in foods by vectors	Andromedotoxins and other substances in honey	Abstention
		Tremetol in milk	Abstention
II	Toxicants concentrated in organisms used for food	Ciguatera poisoning in fish	Abstention
		Saxitoxin in shellfish	Abstention
III	Toxicants occurring normally in plants and animals used for food		
	(i) at all times		
	(a) rapid action	Cyanide in bitter cassava	Leaching and fermentation; selective breeding
		Cyanide in Lima beans	Limited use in a mixed diet; selective breeding
		Cyanide in apricot kernels	Very limited use
		Alkaloids in bitter lupins	Leaching; selective breeding
		Vicine in broad beans	Use in a mixed diet
		Oxalyl substituted amino acid in chick peas	Use in a mixed diet
		Oxalic acid in rhubarb and spinach	Use in a mixed diet
		Nitrates in spinach	Use in a mixed diet

(b) slow acting	Goitrogens in brassicas	Iodine supplementation
	Carcinogens in bracken	Abstention
(ii) only at certain times	Puffer fish poison in fugu	Great care in preparaton
	Hypoglycin in akee	Eat ripe fruit only
	Solanine in potatoes	Avoid peel, eyes, shoots and 'greened' tubers
IV Toxicants formed by interaction of food constituents with (i) other food constituents	Nitrate in vegetables via nitrite to form nitrosamines with secondary amines	Use in a mixed diet
	Nitrite from certain preserved foods with secondary amines to form nitrosamines	Use in moderation and in mixed diets
	Anti-vitamins with vitamins to destroy the specific protective effects	Eating a normal varied diet
(ii) medically prescribed drugs	Active amines in cheese, chocolate, etc. with monoamine oxidase inhibitors	Abstention under medical advice
V Allergens	Idiosyncratic reaction of individuals to certain foods	Abstention based on experience and medical advice

2. Some foods are dangerous in the long term and the recognition of their potential for harm is quite recent.
3. Many foods contain substances which of themselves are poisonous but which in the concentrations found may be safely ignored provided the diet is suitably mixed.
4. Not many of the naturally occurring toxicants pose any threat to the people of the developed nations who enjoy a varied diet. Not all diets in these countries are varied, however, and some groups, in their somewhat irrational desire to be 'natural', run the risk of toxicological danger as well as nutritional imbalance.
5. Local food preferences sometimes include foods which in themselves are hazardous.
6. Those most at risk are populations at the subsistence level who, at certain times of the year, are forced to rely on a particular plant as a staple food.
7. In a modern community the chances of adverse reactions between food constituents and medication are very real.
8. Food allergies are no respecter of persons but sufferers must work out their own salvations.
9. A great deal has been done by the selective breeding of plants to remove many of the hazards which occur naturally in them.
10. Finally, it may be said with some confidence that hazards not yet recognized in food but probably of long-term effect remain to be identified.

3 Chemical Contaminants

'Tis now the very witching time of night;
When churchyards yawn and hell itself breathes out
Contagion to this world.

Shakespeare, *Hamlet,* Act III, Scene ii

In 1956 an unknown disease of the central nervous system was observed at Minamata in Japan. Fish, animals and people died and, in due course, it was found that the cause was mercury poisoning deriving from a chemical factory effluent which flowed into Minamata Bay, contaminated the fish and thus entered the human food chain. The high mercury content of the fish, 10, 20 and even up to 40mg/kg (p.p.m.), and the high fish content of the Japanese diet directly caused the deaths and misery of the Minamata disaster, but when others, alerted by the Japanese experience, began to look elsewhere similar unexpected contamination was found; in Canada, for example, and in America and in Sweden, but under vastly different social conditions. There was concern and fishing was prohibited in certain areas, but there was no outbreak of the so-called Minamata disease. The mercury story is, however, a classic example of the contamination of human food by a deleterious substance, in this case a heavy metal, at dangerous concentrations.

In 1962 Rachel Carson sounded a further warning about contamination, this time about pesticides. Her book, *Silent Spring,* was emotive and exaggerated, but it was salutary because in some places exaggerated things were being done with pesticides. In certain areas men were seeking to exterminate pests of economic significance but with them they were killing off harmless and necessary wild creatures and contaminating human food supplies. Sir Julian Huxley in his preface to the English edition of this book acknowledged the important place of agricultural chemicals but emphasized that the key word was *control:* control of pests, not extermination, and control of exterminators and, we should add, control of the inevitable residues found in food.

In 1973 there was a packaging scare when it was found, first, that a few workers in the industry making polyvinyl chloride (PVC) were suffering from angiosarcoma, a rare form of cancer, attributed to the chronic inhalation of vinyl chloride monomer (VCM) from which PVC is made, and, secondly, that traces of VCM remaining in PVC could find their way in extremely low concentrations into foods packaged in PVC containers. There was no evidence whatever of any harm to anyone from VCM from this source but, again, controls were introduced and very stringent ones they are.

These three examples are all of different kinds and degree of contamination. They have different causes and different effects but all three are food-related and all three would have remained unexplained and uncontrolled without the enormous post-war advances in analytical methodology and instrumentation which have made possible the detection and measurement of extraordinarily low concentrations of complex substances.

Food may be contaminated by chemicals from a number of sources, some from more than one source. Thus, metals have got into food from natural waters, from polluted waters, from pesticides, from the air, from factory equipment and from the final package. Pesticides, also, it could be argued, are industrial contaminants, but these are used deliberately for a specific purpose and are in a different category from chemicals which escape into the environment or are deliberately dumped. Table 3.1 is, therefore, simply an indication of the major chemical contaminants of food, excluding those of microbial origin, and a convenient set of headings under which to discuss them.

Table 3.1: Major Chemical Contaminants of Food

Heavy metals
Industrial contaminants
Residues of agricultural chemicals
Substances formed during cooking and processing
Substances absorbed during processing
Packaging contaminants

Heavy Metals

Mercury: The Minamata disaster, which has cost the company concerned tens of millions of dollars, was followed by other outbreaks of mercury poisoning. In 1960 in Iraq there was an epidemic of mercury poisoning following the use of seed wheat for bread. The seed had been treated with a mercurial fungicide and about one thousand people became sick; 30 to 40 per cent died. The same thing happened in Guatemala in the period 1963-65 when 21 adults and 24 children were affected and 20 died. The worst, however, was again in Iraq in 1972. Yet again, flour from treated wheat was used for breadmaking and 6530 people were admitted to hospital; 459 died from acute mercury poisoning, and the chronic effects are no doubt still visible. In the meantime, Japan had suffered once more.

At Niigata, in the north of Japan, an outbreak similar to the Minamata affair occurred in the period 1965-70. The circumstances were the same; a chemical factory contaminated a river and the victims were fisherfolk. Forty-seven cases and six deaths were recorded.

A singular aspect of the mercury problem is that the metal is changed by microbial and other biological activity into a particular organic form, methyl mercury, which is more readily absorbed than the inorganic form, moves round the body easily and tends to concentrate in brain tissue. About 90 per cent of the mercury in fish is present as methyl mercury and the forms used as fungicides to protect seeds also are organic forms.

All the episodes referred to above were associated with gross contamination of food and high intakes of mercury, 200μg/kg (0.2 p.p.m.)/day for periods of 1-2 months in Iraq, 30μg/kg (0.03 p.p.m.)/day for months or years at Niigata, but scientists involved in public health began to look further afield; to look for mercury in food, all foods, with the sensitive methods then but recently available to them, to examine fish-eating populations for any early signs of mercury poisoning and to study blood and hair mercury levels in high and low fish-eating groups.

Some early studies were done on fishing populations in Sweden, American Samoa, Peru and Sardinia, people consuming, possibly for many years, up to 5μg/kg/day of methyl mercury. The results suggested a low probability of neurological symptoms in individuals with methyl mercury blood concentrations of 200-400μg/l (0.2-0.4 p.p.m.). However, applying a ten-fold safety factor, the WHO recommendation is to aim at a blood level of no more than 20μg/l.

It was soon found that, although mercury is present in foods other than fish and shellfish at very low concentrations (see Table 3.2), it is very widely distributed. It is a normal constituent of sea-water at concentrations of 10-30 ng/l, i.e. 10-30 parts per thousand million; it may occur in fresh water at higher concentrations, particularly in mercuriferous areas, and

Table 3.2: Mercury in Selected Foods*

Food	*National mean* (mg/kg wet weight)	*Maximum found*
Bread	0.003	0.011
Rice	0.003	0.020
Rice, brown	0.004	0.011
Muesli	0.004	0.010
Milk	0.004	0.020
Butter	0.003	0.015
Cheese	0.003	0.015
Skim milk powder	0.003	0.008
Ice-cream	0.002	0.020
Margarine	0.005	0.029
Minced beef steak	0.002	0.005
Lamb chop	0.002	0.005
Pork chop	0.005	0.048
Lamb's fry	0.009	0.025
Poultry	0.006	0.023
Eggs	0.006	0.017
Carrots	0.003	0.008
Lettuce	0.004	0.012
Silver beet	0.004	0.012
Frozen green peas	0.003	0.010
Canned asparagus cuts	0.004	0.014
Orange juice, chilled	0.004	0.011
Mixed dried fruits	0.005	0.016
Prunes	0.003	0.010
Peanut butter	0.006	0.027
Honey	0.004	0.011
Tea	0.002	0.008
Coffee	0.003	0.014
Soft drink, carbonated	0.003	0.008
Beer	0.003	0.014
Wine, red	0.003	0.005

*(Figures taken from The Australian National Health and Medical Research Council *Market Basket (Noxious Substances) Survey – 1976* and *1977*)

Table 3.3: Some Comparative Values of Mercury in Fish from Different Sources

	Number	*Mean*	*Range*
U.K., Freshwater Fish			
Pike	25	0.52	0.06–1.30
Brown Trout	167	0.09	0.02–0.26
Rainbow Trout (fish farms)	29	0.03	0.02–0.08
Sea Trout	47	0.04	0.01–0.12
Eel	40	0.16	0.07–0.30
Tench	9	0.28	0.19–0.37
U.S.A., Marine Fish			
Blue Fin Tuna		0.21–0.68	0.04–0.91
Blue Marlin		3.57–4.78	0.35–14.0
Mackerels		0.11–0.31	0.00–1.31
Sharks		0.46–0.62	0.06–1.56
Flounders and Soles		0.09–0.24	0.00–0.50
Halibut		0.25	0.02–1.52
Cod		0.17	0.04–0.40
Australia, Marine Fish			
Flounder	54	0.04	0.02–0.05
Gemfish	237	0.66	0.07–3.07
Mackerel	77	0.11	0.04–0.25
Marlin, Black	42	7.27	0.50–16.5
Mullet, Sea	84	0.02	0.01–0.20
Shark, Gummy	507	0.44	0.03–3.04
Shark, School	361	0.75	0.01–3.30
Tuna, Southern Blue Fin	219	0.22	0.06–6.63

(Sources: U.K. Working Party on the Monitoring of Foodstuffs for Mercury and Other Metals; E. Zook *et al. J. Agric. Food Chem.*, 24(1):47-53, 1976; *Report on Mercury in Fish and Fish Products,* Working Group on Mercury in Fish, Canberra, Australian Government Publishing Service, 1980, p.133)

it is found in soil up to 100ng/kg, i.e. 1 part in ten thousand million. In the oceans it concentrates in the food chain and is highest in fish at the top of the chain – sharks, game fish, tuna, mackerel – in other words, in fish which eat other fish, and the larger, the older and the more active the fish, the higher the concentration of mercury in its tissues. Some comparative figures are given in Table 3.3 and analyses of specimens held for many years in museums suggest strongly that the mercury content of fish away from contaminated waters is the same as it has always been. Nevertheless, mercury is eliminated from the body only slowly and will build up in the tissues if ingested regularly.

This new awareness, dating essentially from the mid-1960s, led to regulatory activity all round the world. In 1967 the Joint Expert Committee on Food Additives of WHO/FAO (JECFA) urged that the use of mercury compounds which could contaminate food should be strongly discouraged and in 1973 the OECD recommended the elimination of mercury from agriculture and the paper and pulp industry and its stringent control elsewhere to reduce to a minimum its escape into the environment. Mercurials have all but disappeared from those industries everywhere but the metal is still an essential part of the chlor-alkali process for making chlorine and caustic soda, and is used in the electrical industry and for making thermometers. It is also used in medicine and in dentistry.

In 1972 JECFA recommended a provisional tolerable weekly intake (PTWI) of 0.3mg of mercury of which only 0.2mg should be methyl mercury. This was an acceptable weekly intake, not a recommended regulatory limit for food. In many countries health authorities began to reduce the limits for mercury, but the desire to protect the consumer led to some unexpected difficulties as the following experience indicates. In 1971 The Australian National Health and Medical Research Council (NHMRC) recommended to the States and Territories that mercury be limited to not more than 0.5 mg/kg in fish and shellfish and not more than 0.03 in all other foods. In the same year the Customs Department imposed a limit of 0.5 on imported fish and in 1972 four of the six states wrote it into their regulations, Queensland following in January 1973. South Australia adopted a provisional limit of 1.0mg/kg and wrote it into law in 1975.

In 1972 some New Zealand shark fillets were seized at an Australian port because they were found to exceed the limit of 0.5 mg/kg of mercury. The New Zealanders were aggrieved because the fish was the same as was freely available in Victoria and the incident led to an investigation of the south-eastern Australian shark fishery. It was found that school shark averaged about 1.0mg/kg and gummy shark less than 0.5, though individual fish exceeded this figure. Accordingly, in September of that year school shark for sale for food were limited to a length of not more than 41 inches (104cm). Two years later a market survey in Melbourne showed that 60 per cent of the fish fillets offered for sale exceeded 0.5mg/kg and in February 1975 the Federal Attorney-General prepared a Consumer Product Safety Standard under the Trade Practices Act to limit the mercury content of fish to not more than 0.5mg/kg. This was not a good move as it proposed to intrude into the difficult and complicated field of food regulation-making a department which lacked experience in it. Fortunately, nothing came of it. In any case, world opinion which was unanimous in the desire to control mercury in fish to a low level was not unanimous on the maximum limit to be adopted.

Table 3.4 sets out the annual per capita fish consumption for a number of countries compared with the respective regulatory limits for mercury. It is clear that many countries adopt a pragmatic approach related, on the one hand, to their relative dependence on fish as food and the natural levels in the fish they need and, on the other hand, to the relatively small place of fish in the national diet. The need for and place of advice to supplement regulations in communities in which individual per capita fish consumption will vary greatly is implicit in the attitude of certain governments. Such advice has already been given in Australia: against eating fish from the Derwent estuary in Tasmania, against eating more than six pieces of shark per week in Western Australia, against more than three meals per week of Port Phillip flathead and against more than one meal per week of big shark, swordfish or marlin.

The wisdom of such advice was confirmed by the release in 1980 of the McGill University Methyl Mercury Study carried out among bands of Cree Indians in northern Quebec. In summer some of these people live largely on lake fish which often exceed 0.5 mg/kg of mercury and go up to 1.5-2.0 for large pike. Results of systematic medical examinations showed association between 'mild' or 'questionable' neurological abnormalities and estimated exposure to methyl mercury and demonstrated that groups eating very large quantities of fish from waters uncontaminated by industry could be at risk. Accordingly, the Grand Council of the Crees in Montreal issued to the various bands advice on how many each of specified varieties of fish from specific lakes or coastal waters might be eaten in a given period in order 'to keep their blood mercury safely below 50ppb (sic)'. That is a figure two and a half times the WHO aim of 20 but very much lower than the 200μg/l said to be the level at which overt symptoms may appear in a few people.

Late in 1971 New Zealand adopted a limit of 0.5mg/kg but some fish exceed this concentration. School shark average more than 0.5 and, while a typical catch of snapper may average, say, 0.26, individual old fish could go as high as 1.3. Trout from the geothermal area have reached 3mg/kg and a survey in Auckland suggested that some 2 to 3 per cent of children in that city had an intake of mercury from fish (shark) and chips high enough to cause some disquiet. In other words, apart from the peculiar circumstances of geothermal activity, the New Zealand experience is little different from the Australian.

For several reasons the mercury story has been told at some length. The effects of mercury were unsuspected; they were dramatized by the con-

Table 3.4: Some International per capita Fish Consumption and Regulatory Limits for Mercury*

Country	*Annual p.c. consumption of fish and shellfish, kg*	*Regulatory limit for mercury, mg/kg*	
Iceland	39.1	1.0	
Japan	36.4	(a) Provisional guideline 0.4, methyl mercury, 0.3 (b) Tolerable weekly intake of methyl mercury 170 μg; less for pregnant women and children (c) Fishing prohibited in some waters (d) Certain fish species with high natural levels exempt	
Denmark	35.5	1.0	
Portugal	22.8	No limit for domestic use	
Spain	17.0	0.5	
Thailand	15.5	0.5	
Finland	13.2	1.0	
Norway	11.5	(a) Provisional tolerable weekly intake of 0.3 mg (0.2 mg methyl mercury) (b) Warning against fish from Sorfjorden	
U.S.S.R.	10.2	River fish	0.2
		Ocean fish	0.3
		Ocean fish and products	0.5
		Fresh tuna	0.5
		Canned tuna	1.0
Korea	9.8	0.5	
Greece	9.1	0.7	
Belgium and Luxembourg	8.2	No limit. Fish significantly exceeding 0.5 sometimes withdrawn from the market	
France	7.9	0.7 for fish expected to be high, otherwise 0.5	

Table 3.4: (continued)

Country	*Annual p.c. consumption of fish and shellfish, kg*	*Regulatory limit for mercury, mg/kg*
Netherlands	6.4	Unofficial guidelines, 1.0 for freshwater fish, 0.5 for marine fish
Canada	5.9	0.5
U.S.A.	5.5	1.0

*(Prepared from data in *Report of Mercury in Fish and Fish Products,* Australian Department of Primary Industry, 1980)

tamination of a food supply by gross industrial pollution; they resulted from very low concentrations in absolute terms; they, and the potential dangers in certain types of food supplies, were revealed only by modern methods of analysis made available in the last thirty years, and they raise the question of what other similar dangers remain to be revealed. In addition this story emphasizes the ability of modern scientific techniques to solve a problem and to monitor a food supply, and demonstrates the difficulties of protecting every individual by regulation alone. Regulation, policing and sensible selection of food on the basis of informed advice on contamination levels all have their place.

Lead: Some people consider lead to be a bigger threat than mercury, mainly because it is far more widely spread in the environment. Certainly, lead poisoning has been known for a very long time. It has been seriously suggested that the aristocracy of imperial Rome suffered from lead poisoning from lead-lined cooking pots and other vessels of pewter (70% tin, 30% lead). And there was the elderly man in Melbourne fifty years ago who developed lead poisoning, the cause of which was a mystery until it emerged that he waited each morning for the pub to open and had the first glass of draught beer which had in those days been lying all night in the lead pipe between the barrel and the tap.

Lead is a cumulative poison. It can reach the foetus and it affects the brain and thus, because of potential brain damage in the late foetal stages and in the first eighteen months of life, it has now become of special concern. Lead is absorbed from food, water, air and, especially by little children, from other things on which they bite and chew and which were once coloured with lead paints. While the effects of high levels of lead intake are well known, there has been much discussion, even controversy, in the 1970s of the effects of continuing exposure to very low levels. The main concern has been over a contention that lower concentrations than those required to produce overt signs of lead poisoning lead to lowered I.Q. and behavioural problems in children. However, a U.K. Department of Health and Social Services Working Party reported in 1980, *inter alia,* that there was no evidence of deleterious effects at blood levels below about 350μg/l and no doubt at all about the ill effects at levels above 800. Between the two, it agreed, there was still uncertainty. This solved nothing.

Lead finds its way into food from water, the soil, the air, fruit sprays and the final container. Hundreds of samples taken along the New South Wales coast (some 700 miles) averaged 0.8mg/kg of lead in oysters and 0.61 in fish, one sample of the latter showing 4.1. Vegetables grown in Derbyshire soils ranging from 200 to 7000mg/kg of lead contained higher than normal, but acceptable, levels of lead which were concentrated in the skins of tubers and roots, the outer leaves of brassicas and the pods of peas.

Food plants grown near lead smelters and metal refineries have given abnormally high values. Leafy vegetables have shown higher contamination than root crops, lead in the former, particularly, falling with distance from the refinery, but by far the greatest proportion of atmospheric lead comes from petrol. In New Zealand sheep grazing near major highways have shown higher blood and tissue lead than controls away from traffic; but English studies have shown air-borne lead to drop sharply with distance from the verge, 20% at 50 metres, 10% at 150 metres. Australian work, on the other hand, provided presumptive evidence of fall-out on vegetables grown in Melbourne suburbs as well as intake from spray residues on fruit.

Cans are of tinned steel or of aluminium. The former are soldered along the side seam with a lead solder, the latter are drawn and have no side seam. It is clear from analytical data that solder is a source of lead and international concern has led to the reduction of the regulatory limit of not more than 2mg/kg for baby food in cans to 0.8 with a further reduction to 0.3. There has thus been a concerted effort by industry worldwide to develop and introduce the technology of making cans with welded not soldered side seams. It has been shown also that many paper products used in packaging food contain lead. In an American study, four samples of paper wrappings for food ranged from 0.6 to 1.0 *per cent* of lead. Other samples contained less but still carried enough lead to cause considerable concern. The metal came mainly from the pigments of coloured printing inks but recycled newspaper is a likely source of lead in a number of paper and fibre-board food wrappers and containers.

Lead salts make very good glazes for ceramics and pottery. Unfortunately, if dishes and hollow-ware with such glazes are used for food and drink some lead is leached out and consumed. This is well recognized and regulations governing such ware intended for food use are in force in many countries. It is important, however, that consumers themselves be aware of the possibilities, ask about the wares they purchase and refrain from using for food, and especially drink, vessels clearly intended for ornament.

Cadmium: Cadmium may be absorbed from food and from tobacco smoke. Although most of it is excreted, 50 to 75 per cent of that portion which is absorbed is stored in liver and kidneys and thereafter is got rid of very slowly. Accordingly, continuing exposure even to very low levels may cause a build-up of cadmium in the body and lead to kidney dysfunction and other symptoms called in Japan Itai-Itai disease.

Cadmium is normally present in soil at concentrations of, say, 0.2-0.6mg/kg. In Japan, where industrial contamination has occurred, levels of 1-69mg/kg have been found but some soil concentrations may be increased by traces of cadmium in fertilizers. From the soil the metal passes into food plants – cereals, brassicas, legumes and root crops. Cadmium differs from lead in that there is a continuing accumulation of it in plants whereas accumulation of lead begins only after a relatively high concentration is reached in the soil. Cadmium differs also from mercury in that there is some evidence that it does not accumulate in the food chain. Nor does it pass the placental barrier and reach the foetus as do both mercury and lead.

Figures for the concentration of cadmium in foods show that most contain less than 50μg/kg with a range of 15-70μg/day reported for Europe, North America and Japan. Cadmium is naturally higher in shellfish and crustaceans than in other foods and in Japan rice up to 1000μg/kg was not uncommon, but this was due to industrial pollution. Concentrations of 0.5-4mg (i.e. 500-4000μg)/kg in shellfish and 10mg/kg for the brown meat of crabs (0.3 for white meat) have been reported. The American oyster, for example, has been shown to accumulate cadmium from aqueous concentrations as low as 0.0005mg/l. An official survey reported in the U.K. in 1973 showed 21 samples of oysters taken between Margate and Harwich to average 1.1mg/kg of cadmium (range 0.36-1.9). This compares with a range of 0.1-1.0mg/kg (mean 0.2) for hundreds of samples of oysters taken in New South Wales in 1975 and 9-40 in those from a polluted waterway in Tasmania. Oysters from unpolluted but highly mineralized areas of Tasmania range up to 7mg/kg, confirming observations in other parts of the world that, in general, the high cadmium in shellfish seems to be of geological rather than industrial origin.

The U.K. survey referred to above suggested that in Britain the only sections of the community which might consistently ingest greater than average amounts of cadmium would be those eating large amounts of limpets, winkles and dog whelks from the Bristol Channel east of a line from Lynton to Worms Head, brown body meat of crabs, or kidney. This is obviously no cause for alarm when compared with an apparent anomaly observed in New Zealand where oysters from Foveaux Strait range up to 8 mg/kg and are eaten very freely in the port of Bluff for six months of the year. Only one dozen per day supplies more than 3 mg of cadmium per week which is six to seven times the provisional tolerable weekly intake (PTWI) of 400-500 μg (0.4-0.5 mg) recommended by WHO but the local health statistics suggest no detrimental effects.

Cadmium is not absorbed from cans but may be picked up from certain glazes on ceramics. Because of this it, like lead, is regulated in glazes on dishes and other containers intended for use with food.

Arsenic: Arsenic is not a heavy metal but it is toxic and it is convenient to discuss it under this general heading. Arsenic has been known for centuries to be poisonous and it has been used in medicine for a long time. It has formed the basis of weedkillers, fungicides etc., has been used, as Paris Green, as a colour in wallpapers and has long been known to concentrate in marine creatures.

Food may pick up arsenic from soil and water and may be contaminated environmentally and by carelessness. Normally, arsenic occurs at very low levels, say, 0.05 mg/kg in plants used for food and at five or six times this concentration in meat. However, the major food sources are fish and shellfish in which levels of 5-6 mg/kg are common and values up to 60 and even 100 are not unknown.

The toxicity of arsenic varies with the form in which it is found, several variations being possible. Inorganic arsenic is the most toxic and the organic form, which makes up most of the arsenic in marine creatures, is poorly absorbed by the body and is excreted relatively rapidly via the kidneys. Surveys of communities consuming large amounts of seafood have shown no adverse effects; consequently, food regulations are tending to differentiate between organic and inorganic arsenic and to limit the latter in fish, crustacea and shellfish to not more than 1 mg/kg.

Inorganic arsenic in drinking water has been reported in several parts of the world. Concentrations approaching and exceeding 1 mg/l have been found in water associated with arsenical ores and geothermal activity and have caused signs of arsenic poisoning in the communities using them.

In New Zealand it has been estimated that geothermal activity puts 64 000 kg of arsenic per annum into the Waikato River system. In spite of this, drinking water is below the WHO recommended limit of 50μg/l. Trout from the area have shown concentrations in equilibrium with the water and, hence, of no consequence, but lakeweed has been found with values up to 1 000 mg/kg (i.e. 0.1%) of arsenic in the inorganic form. This finding must raise some doubts about dried kelp, the so-called health food supplement, for which concentrations of 20 mg/kg are known.

Organo-arsenic compounds have been used as growth promoters for poultry and animals used for food and concentrations of more than 1 mg/kg have been found in, for example, chicken livers. It seems hard to justify a continuation of this practice.

Tin: Tin finds its way into foods from the cans in which food products are so frequently packed. The tinned steel can dates from 1810 and is still the most commonly used food package. While the tin protected the food from a very large pick-up of iron, tin itself was dissolved in relatively large amounts so that for many years the maximum permitted limit for tin in canned foods was 250 mg/kg and this was by no means too high for fruits and other acid products. Corrosion and staining of the can surface led to the development and introduction of internal lacquers for cans and a reduction in the tin uptake so that the permissible limit for tin has easily been reduced to 125 mg/kg.

Beyond the fact that the body can tolerate very much more tin than mercury, lead, cadmium or arsenic not a great deal is known about this metal. The amount of information about tin in foods, apart from canned foods, is low but there is a report of fish from an uncontaminated Canadian lake containing 2.6-5.4 mg/kg, presumably from geological sources.

Other Metals: Iron, copper, zinc, cobalt, nickel, chromium, manganese and selenium are trace elements actually required by the body for various purposes. Some of them, such as selenium and zinc, are said to reduce the toxicity of mercury and cadmium respectively but they are all toxic in excess.

It is well known that selenium may be absorbed into cereals grown on selenium-containing soils and it is possible for unacceptable and dangerous concentrations of all these elements to be absorbed by creatures used for food, especially fish, as the result of industrial pollution. This has been amply demonstrated by detailed studies of the Derwent estuary in Tasmania in which concentrations (mg/kg) in fish up to the following were found: chromium, 0.3, cobalt, 1.6, copper, 5.3, manganese, 4.0, nickel, 1.5, and zinc, 18.0. In shellfish the concentrations were even higher: chromium, 3.2, cobalt, 2.6, copper, 296, manganese, 95, nickel, 4.3 and zinc, 8580. An earlier study recorded concentrations of zinc in oysters far exceeding those required to cause vomiting in man. In one, 10 per cent of the dry weight was zinc, a world record. This pollution, coming from an electrolytic zinc recovery plant, destroyed an attempt to establish an oyster farm, and fish and seafood from the Derwent waterway must remain suspect.

In 1977 a working group of the Australian Academy of Science on monitoring noxious substances in foods reported, *inter alia*:

> So far there is no evidence of any toxic effect on Australian consumers from the ingestion of heavy metals in foods, but the working group wishes to stress the need for continuous monitoring of the levels of these potentially toxic substances in food, in view of the growing industrialization and urbanization of the community and the life-time exposure that they present. (Australian Academy of Science Report No. 22, *Food Quality in Australia,* 1977, p. 12).

It went on to point to the interactions of metals with each other and with other food constituents and its views reflected those expressed at much the same time by U.K. working parties and in other countries. It is evident that all over the world research on heavy metal contamination and the means of preventing it is continuing at a high level and that, in the light of much new knowledge, allowable limits for heavy metals in foods are regularly being reduced.

Industrial Contaminants

The mercury poisoning at Minamata and Niigata and the cadmium poisoning of Itai-Itai disease resulted from industrial contamination, but the intense research activity which the identification of the causes of these disasters initiated revealed that naturally occurring mercury, cadmium and other metals could appear in certain foods under certain conditions in potentially hazardous concentrations.

The contamination of the Derwent estuary in Tasmania is industrial in origin, so is the potential contamination of home-grown vegetables in the Hobart area and of food from plants grown anywhere in the world in the vicinity or leeward of industrial activity of many kinds.

To this extent then, we have already been discussing industrial contamination but there remain many thousands of chemicals used in industry which, if not disposed of safely as products or wastes, or if allowed to escape, have the potential seriously to contaminate food. Industrial contamination is, therefore, of two kinds:

1. Contamination of the sources of food by which the food, animal or vegetable, is contaminated as it grows.
2. Accidental contamination of the food itself.

Of these the former is more prevalent and more likely and the metal contamination already discussed belongs to this class. So, too, does contamination resulting from the dumping of wastes or uncontrolled effluents, though some of the latter, while unpleasant, threaten no chemical contamination of food; whey from cheese factories and wastes from fruit canneries are examples.

The necessity for the control of effluents is now well understood and tight controls are imposed by the environment protection authorities. The dumping of wastes, too, is known to pose problems as experience in many places has only too well shown. The polychlorinated biphenyls are an excellent example.

These components, known as PCBs, are a group of chlorinated hydrocarbons first produced commercially in 1929. They are used extensively as dielectric fluids in the electrical industry, as heat exchange fluids, as plasticizers, in inks, in marine anti-fouling paints, in textile coatings, in some pesticide formulations and even in kiss-proof lipsticks. Carbonless copy papers contained up to 3 per cent and hence PCBs appeared in food packaging materials made from recycled paper. The accumulation of PCBs in the environment, and hence their potential appearance in foods, was first demonstrated in 1966 and their detection almost anywhere in the world is now possible. The commercially desirable property of PCBs, and the one which guarantees their persistence in the environment, is their chemical inertness. They are hard to dipose of and ordinary incineration simply serves to vaporize them into the atmosphere. Temperatures of at least 1100°C are necessary to destroy them.

The acute toxicity of PCBs is low but chronic toxicity may be a different matter. PCBs are very soluble in fat and are therefore prone to be absorbed by fatty foods and, like DDT, to accumulate in body fat. In 1968 a thousand people in Japan were affected by PCBs when a batch of rice oil was contaminated with 1000-1500mg/kg (parts per million) of these compounds by a leakage from a heat exchanger. Pale yellow cysts were found throughout the body and neurological and respiratory symptoms appeared. Those affected were thought to have ingested up to 2gm of PCBs whereas the minimum amount to produce symptoms was put at 0.5g. Do PCBs produce cancer? We don't know because they may contain as impurities other substances which do and may form yet others when heated to high temperatures but it is clear that there is every reason for tight controls on the presence of these compounds in the environment and especially in foods.

In America there have been cases of gross contamination of rivers and soil with PCBs from improper waste disposal and industrial accidents, but now PCBs are limited to closed systems and banned from food and feed plants. Their use in copy paper has been discontinued and recycled paper containing copy paper is no longer used for packaging materials.

In 1976 the U.S. Congress passed a law phasing out the production of

PCBs, but there remained two questions. What should, or could, be done about the PCBs and PCB-containing equipment, etc. already in being? And what about PCBs in the environment and hence in the food chain? PCBs are very stable and persistent. Destruction is by incineration at 1100°C and is thus a specialized undertaking. Because of this there are few disposers but regulations promulgated originally in 1979 and reviewed in 1982 put a time limit of one year on the storage of PCBs and PCB containers prior to disposal and a 90 day limit, within that year, between receipt by the disposer and actual disposal. In 1982 it was estimated that there were 17 million kg of PCBs in Canada. Most of them were still in service. No new PCBs are permitted to be used and all PCB-containing equipment is to be phased out over the next 40 years. In the meantime, there are strict guidelines on everything containing 50p.p.m. or more of PCBs. These compounds are well established in the environment, especially the aquatic environment, and will persist for a long time. They may be found in food at concentrations low enough to cause no concern. Limits equivalent to the analytical limits of detection seem appropriate, the exception being fish which, both as a commodity and as sporting trophies, has been studied intensively by the U.S. Food and Drug Administration.

This episode is a good example of:

1. The contamination of food by a chemical serving very useful technological purposes but harmful in the body;
2. The control of this chemical *vis-à-vis* the food supply;
3. The necessity for the continuous monitoring of all industrial chemicals on the assumption that sooner or later, by one route or another, they will find their way into food.

Many other industrial contaminants are chlorinated hydrocarbons; traces of dioxin as impurities in polychlorinated phenols and 2,4,5-T, chlorinated phenols and chlorinated benzenes. After the disaster at Seveso in Italy in which dioxin escaped from a chemical plant, this substance was found at low levels in milk from farms in the area. Other polychlorinated hydrocarbons have been identified as entering the food supply from the fly ash of incinerators but it must be emphasized that most of such knowledge now available is due entirely to the ability of modern analytical techniques to determine this class of substance at concentrations of parts per thousand million. They should not be there at all but the significance of such low concentrations in terms of public health is not known.

All food manufacturers live with the knowledge that accidental contamination of their raw materials or final product is possible. Precautions to guard against it are, therefore, built into their technology. Two recent examples of what can happen came from the U.S.A. In the first, animal fats were contaminated by PCB which escaped from a damaged transformer and the second involved polybrominated biphenyl (PPB), a first cousin of PCB. A fire retardant containing PBB was sent out to an animal feed plant in Michigan instead of feed supplement and became part of some batches of stock feed. It took almost a year to identify the cause of the resulting rather gruesome animal poisoning by which time PBB had become well established in local human food supplies and is likely to remain as a low-level contaminant for a long time. The relationship of PBB to human health problems in the area is not fully known and will remain an unresolved question for years.

Residues of Agricultural Chemicals

Agricultural chemicals include fertilizers, weedicides, insecticides, and fungicides, but of these the most usual food contaminants are the last two, collectively referred to as pesticides, which are used to protect animals, fruit, vegetables and other crops in the field from all kinds of insects and from attack by moulds, viruses and mites. They are also used to prevent the growth of molluscs, nematodes, plants or animals which may cause serious economic losses of food by destruction of it in the field and in store and, in the consumer societies, by rejection of blemished fruit in the market place. Though such crops as wheat rely on selective breeding for protection against fungus diseases such as rust or smut in the field, insecticides must be used in silos, storage sheds and ships to keep insects at bay. Cool storage of fruits and vegetables is itself a protection as insects are most unhappy at temperatures below 15 °C. Although there seems little doubt that the prodigal use of pesticides has adversely affected some species of fauna, their value in conserving and improving the world's food supplies cannot be challenged.

DDT and other organo-chlorine compounds have been in use now for decades. They are widely distributed throughout the world and doubtless we all carry traces of them in our body fats. In Italy during the war millions of people were dusted with DDT powder to stop an incipient typhus epidemic. Others have lived for years in close association with DDT and volunteers have been dosed to yield levels of 200-300mg of DDT per kg of body fat all without any signs of disease or harm. DDT is probably the safest of the insecticides but it was made the scapegoat for what was by any standards the massive over-use of insecticides in North America twenty years ago. Its value to the world has, however, been enormous.

Other insecticides include organo-phosphates, which are readily degraded into generally unobjectionable substances, carbamates and quaternaries. Organic mercurials have been used as fungicides on seed grain and the disastrous effect of using such grain for food has already been discussed. Such episodes are, however, rather different from what may be expected from contamination of commodities intended for consumption. Weedicides such as the substituted ureas, Monuron and Diuron, and the chlorinated phenoxy-acetic acids (2,4-D and 2,4,5-T) are also used in agriculture and horticulture.

Because of the adverse effects of DDT and other compounds on various fauna there has been great debate on whether or not DDT should be banned. Among others, a specialist committee set up by the Australian Academy of Science in 1971 studied the matter in depth. It concluded, *inter alia,* that most of the DDT used in Australia was used on non-food crops such as cotton and tobacco, that the regulation of DDT there had resulted in less abuse than in many other countries, and that the levels of DDT and its derivatives in foods for local and export markets were well within permissible international limits. Analyses confirmed the presence of DDT in Australian wildlife but the overall effects were slight and nothing like those associated overseas with the massive use of the insecticide.

DDT has not been banned in Australia but it is thoroughly understood that residues will remain on food, will be metabolized by the body and therefore must be strictly controlled. This, of course, applies to all insecticides and for each of them Australian food regulations list the specific foods in which they are permitted to appear and the maximum amounts in mg/kg allowed. The latter range from 0.000006 mg (i.e. 6 ng) for Ethoprophos in water to 50 for Capran in celery and carbon tetrachloride and ethylene dichloride in raw cereals. The fumigants, hydrocyanic acid and bromide (from methyl bromide), are permitted at 75 in raw cereals and 400 in spices and herbs respectively. New Zealand has similar, but not identical, schedules.

It is now well known that strains of insects resistant to DDT eventually emerged and to combat them other classes of insecticides were developed successively. As further resistance built up, still more insecticides were introduced and most of them were more toxic to mammals than DDT.

It has become obvious that it is not possible to exterminate a species of insect any more than it is possible to exterminate an idea – the idea of the control of insects by insecticides, for example. But the application of insecticides was too much of an empirical hit or miss affair and only recently has a viable theory of why insecticides work been put forward. On the basis of it the CSIRO Division of Applied Organic Chemistry in Melbourne has prepared a series of compounds, called insecticidal esters, which are undergoing field trials. They are most effective against insects and have a very low mammalian toxicity. In another study altogether the CSIRO Division of Mechanical Engineering has had a good deal of success on a large scale in controlling insects in chemical-free wheat by holding silos and storage sheds at air temperatures less than 10 °C. In addition, work is well advanced on an in-line system of heat treatment aimed at destroying insect life at the dockside as the wheat is delivered into the ships' holds.

The first of these advances holds out the hope of properly formulated highly effective insecticides of very low toxicity to all mammals including, of course, man and, specifically, reduction of risks from residues in foods. The second suggests that one major food commodity may be thoroughly protected in store and reach the market without the use of any insecticide at all. A third method, referred to in Chapter 4, is irradiation.

Biological control of insects has been practised for some time and there are those who promote it to the exclusion of all else. Its application is no more free from problems than any other method and it seems that the concept of pest management including chemical, physical and biological methods, is the correct one. If this is so there will always be chemical agents, i.e. pesticides, and, therefore, pesticide contamination of foods. It follows inevitably that there will always be the need for the regulation and monitoring of these residues.

Finally, it is worth noting that food processing is often able to reduce the pesticide concentrations in the finished product. There is, for example, an obvious apparent discrepancy between the new limit of 5 mg/kg for hydrocyanic acid (HCN) in marzipan (see Chapter 2) and the limit of 75 permitted in raw cereals after fumigation. However, during the great aeration of milling and the heating of baking etc., the HCN is almost entirely dissipated. As may be expected, washing, especially if multistage, reduces contamination considerably. So does blanching, which is a preliminary mild heat treatment with steam or hot water. Peeling, where that is possible, is particularly effective and some pesticide residues are degraded during cooking and heat processing. The ordinary unit processes of the food manufacturer's plant are therefore working in the direction of reducing pesticide residues and the consumer is well advised to wash leafy vegetables well and, where it is possible, to peel fruit before eating it. Grapes and cherries pose problems but they can at least be washed.

The foregoing discussion has been concerned almost exclusively with the use and effect of insecticides on plant products but the treatment of animals may also lead to residues of a different kind. Concern has been expressed in America on the contamination of meat and animal products such as milk, butter and cheese by the spraying of cattle feed with pesticides but this is not a problem in countries where livestock graze all the year round and lot feeding is at a minimum. Animals are, however, dipped to combat ectoparasites such as ticks, and treated with drenches and other stock medicines for a variety of reasons.

The treatment of mastitis in dairy cattle with antibiotics, especially penicillin, leads to residues of the antibiotic in the milk. These have induced violent reactions in people sensitive to penicillin and have stopped the lactic starter organisms used in the manufacture of cheese and yoghurt.

Anthelmintics (for the treatment of worms) also may appear in the milk, and other biologicals, administered by injection as well as orally and intra-mammary, may leave residues in human food. This is well known and, as for penicillin, there are stipulated time intervals between cessation of treatment and the use of milk products for human consumption or the slaughter of the animal for food.

The practice of increasing the body weight of various food animals by feeding or implanting certain hormones, specifically oestrogens, was used quite extensively at one time in the United States, with obvious benefit to the producer. Too little was known about the possible effect on man of traces of such hormones which could be transmitted by meat, and their use was forbidden in some countries before it ever began.

Substances formed during Cooking and Processing

Foods are mixtures of a great many chemical compounds so it is not surprising that when they are cooked, and heat processing is simply cooking in the can or jar, chemical reactions occur. Indeed, cooking is intended to promote those reactions on which palatability and digestibility depend, but other less desirable reactions also take place.

Boiling, steaming and even pressure cooking of food are carried out at relatively low temperatures at which only beneficial reactions occur. Roasting, however, and especially barbecuing, subject localized areas of food, especially meat, to high temperatures which frequently produce incipient carbonization and lead to the formation of known carcinogens. Over an open fire the fat is pyrolyzed and absorbs smoke which is often dry-distilled from wood and contains some most unpleasant substances. Hot-plate cooking over embers is far safer, but even this and oven-roasted meats have been shown to contain what would be regarded in other foods as contamination.

Other normally accepted culinary activities designed to produce desirable flavours also produce substances which are chemically dubious. The toasting of bread, the browning of cakes and other baked goods, the production of caramel sauce by heating a can of sweetened condensed milk in boiling water all depend for their flavour and colour on a reaction between proteins and sugars. Called the Maillard (after the discoverer) or browning reaction, it is well known in food science, and important in food technology for the production of a number of desirable flavours and sometimes for the unwanted production of carbon dioxide, but it also produces traces of some substances which would be highly undesirable at higher concentrations and it may interfere with the absorption of some proteins.

This reaction, or series of reactions as it really is, also may take place during food processing such as in baking, as noted, in the alkali treatment of proteins and in the roasting of coffee and coffee substitutes in which carcinogens also are known to be formed.

Nitrosamines (see pp 29, 80) are formed during the frying of bacon and some extraordinary precautions, such as opening the windows and standing back from the stove, have been advised by consumer groups in America. Such suggestions are bizarre. Nitrosamines may also be formed during the curing of meat from the use of nitrates and nitrites as additives (see p. 80).

Fats and oils are refined, bleached, deodorized and hydrogenated in all of which there are possibilities for the formation of potentially toxic compounds. However, there is more potential harm in the oxidation, hydrolysis and polymerization of fats in those fish shops and 'take-aways' which merely top up their oil baths instead of recharging them, say, weekly.

Some years ago, it was found that the bleaching of flour with nitrogen trichloride formed a compound which caused running fits in dogs. The use of this substance was at once discontinued but the incident led to a continuing watch on the possible effects of other bleaching agents. Traces of products formed by reaction with solvents and fumigants have also been detected and feeding tests at high levels on experimental animals have suggested some toxicity; but what happens in man, if anything, is not known.

Alkalis are used in food processing to peel fruit and vegetables, to dissolve proteins for the manufacture of concentrates, isolates and spun fibres and to produce proteins with special, say, foaming, properties. Alkalis have been shown to produce with proteins a substance toxic to rats, but, again, its significance, if any, in public health is doubtful.

Fermentation for wine and spirits produces ethyl alcohol, a toxic substance which is acceptable, but if conditions are not controlled alcohols more toxic than ethyl alcohol (and methyl alcohol is the most common) and other groups of compounds, such as aldehydes, may be formed with fatal results. Similarly, uncontrolled vinegar fermentations may lead to the formation and accumulation of undesirable compounds.

Most food products call for temperature and other conditions too mild for the formation of hazardous compounds and those mentioned above are found in very low concentrations. That they are found at all is not surprising given the complexity of even simple food systems. Many have been present in food since cooking began or since man first commenced to bake and ferment foods. Others result from relatively modern practices and whether or not they have any significance for man is still uncertain in most cases. This very uncertainty suggests a low level of concern but research continues in all these fields and, as they are obtained, regulatory authorities gather the results for assessment.

Substances absorbed during Processing

Water, steam, the equipment itself and detergents and sanitizers used to clean the equipment both chemically and microbiologically are all potential sources of contamination of food. Water will contain traces of dissolved solids naturally present and the concentrations of these will vary greatly from place to place. Water may also introduce micro-organisms and is therefore usually chlorinated to prevent this. Chlorination itself results in the formation of traces of organo-chlorine compounds, called trihalomethanes, some of which are known to be carcinogenic. The concentrations in water, however, are fractions of nanograms/l and the dangers of using unchlorinated water are much greater than those from such low concentrations of these contaminants. The latter may, however, introduce flavour taints.

Steam is a possible source of contamination. Scale may gradually build up on boiler tubes and reduce efficiency, and certain organic and inorganic compounds are added to the water to prevent this. If volatile they get into some foods or they may be carried over in the steam as tiny droplets, but, because these possibilities are recognized, those substances proposed for the food industry are treated as potential additives for the use of which permission must be obtained in the usual way (see Chapter 7).

Contaminants from the equipment are essentially metallic and may find their way into the products by solution and abrasion. This has already been implied in the early part of this chapter. A couple of examples will suffice to show what did and could happen. The first concerns a flavour defect. For many decades the dairy industry used tinned copper equipment. Over a period the tin coating wore off and milk was frequently exposed to bright copper surfaces with the result that there was a pick-up of copper which carried through into the various products manufactured from the milk. Copper is a strong catalyst of oxidation and its occurrence in dairy products such as cheese at concentrations higher than about 2 mg/kg leads to unpleasant flavours. The copper contamination posed no public health problem but the flavours induced by its presence were a serious defect.

In the second example iron was the contaminant, but again there was no question of safety. About forty years ago a well-known jam manufacturer received a complaint from an old lady that a sample of the company's apple jelly turned her teeth black. The offending sample was retrieved and was found to have a slightly metallic flavour. Chemical examination revealed a high level of iron. It did not take a great deal of insight to picture the complainant drinking tea with her bread, butter and apple jelly, and, by reaction of the iron with the normal tannins of the tea, producing what was, in effect, a mouthful of ink; a surmise which was quickly verified by experiment. The product was being made with the normal pre-war mild steel equipment from which the naturally acidic apple juice from which the jelly was made dissolved an appreciable amount of iron.

Metal contamination has been virtually eliminated from the food processing industry by the universal adoption of stainless steel equipment. At the same time, plastic components have been introduced and the known transfer of various substances, especially monomers and plasticizers, from the plastics to the food has led to the establishment of standards for all plastics for food contact (see pp 64-5).

Cleanliness has always been of first importance in food processing. Visible dirt or extraneous matter (what the Americans dramatically call 'filth') in food is bad both for the manufacturer and consumer and the build-up of scale or product debris on equipment reduces plant efficiency and provides reservoirs for contamination by unwanted micro-organisms which lead to food spoilage or public health problems. Accordingly, detergents for removal of soil and sanitizers for killing micro-organisms are part of normal plant sanitation. In many food plants cleaning-in-place (CIP) systems have been worked out to enable pipelines and associated vats, pasteurizers and other equipment to be cleaned and sanitized without having to dismantle them. These systems are being computerized and their introduction has required the development of detergents and sanitizing agents with special properties specifically for the food industry. They are therefore non-toxic in the normally accepted sense of the word; it is presupposed, however, that all equipment and surfaces treated with them will be thoroughly rinsed with clean water before re-use. There is no problem with a computerized factory CIP system, but on a dairy farm or in a non-automated factory it is possible for detergent and sanitizer solutions to lie in the pipelines and be swept into the milk or first batch of the product. This danger was brought home to everyone by the iodine-in-milk episode which serves as a useful example of this kind of contamination.

In the early 1970s questions were raised in the scientific literature about the possibility of increased amounts of iodine finding their way into the diet from various contaminating sources and cases of thyrotoxicosis were identified in Launceston in Tasmania. That island had long been known as a goitre area but iodine supplements had been used in bread and salt to correct this. Now, iodine itself is a goitrogen and it was not long before those cases were associated with increased intakes of iodine from the use in dairies of a group of iodine-containing compounds, called iodophors. Because of their effectiveness in cleaning up bacterial contamination in dairy equipment, these compounds had been introduced to the dairy industry by the various Australian state Departments of Agriculture. Without realizing the implications, farmers and dairy factories used them freely even to the extent of adding them to farm milk to reduce bacterial counts. There is no doubt that their legitimate use reduced bacterial contamination greatly but analyses in all states revealed levels of iodine up to 3000 μg/l of milk. Concentrations of 500-1000 were quite usual whereas normal uncontaminated milk would be expected to contain, say, 30-100. Milk products such as yoghurt and ice-cream also contained raised concentrations of iodine.

No one really knew what it all meant in public health terms, but everyone knew that the smallest people, infants, drank the most milk, weight for weight, and prompt action was taken through Departments of Agriculture to reduce iodophor usage to levels of good manufacturing practice. It was recognized that iodophors were valuable in improving the quality of the milk supply and were less objectionable than some other substances also in use, but control of their use on farm and in factory was essential and was backed up by a regulation limiting iodine to 500 μg (i.e. 0.5 mg) per kg of food.

Packaging Contaminants

Food is packaged to protect it from contamination by micro-organisms, dust, insects, rodents and so on, but it is inevitable that substances of various kinds from the package itself will find their way into the food. Because they migrate from the package to the food they are referred to collectively as packaging migrants. Over the centuries all kinds of things including organic chemicals, sulphites and heavy metals would have been picked up by foods and beverages from barrels, with or without pitch linings, and from jars, bowls, pots and dishes made from wood, ceramics or metals. The same is true today but under modern conditions much more food is packaged for transport and storage and the type of package will depend on the type of product. In general, however, it will come from one or other of the following groups: metals, cellulose, glass and ceramics, plastics.

Metals: The most common package is the can which consists of a very thin layer of tin on a steel base plate. Cans are made up usually from lacquered plate which will protect the inside of the can from contact with the food. Nevertheless, depending on the type of food and length of storage, there may be pick-up of tin, lead from the solder, and even iron from the base plate if for any reason lacquer or tin coatings are penetrated.

The first cans were made by hand from sheet tin plate soldered down the side seam and top and bottom with a lead/tin, mostly lead, solder. Though the technology of can manufacture changed radically, the soldered side seam remained and reference has already been made to the concern, and the reasons for it, over the contamination of canned foods, particularly baby foods, by lead from that seam. Reference has also been made to the limit of not more than 0.3 mg of lead per kg of those products. Results freely available in the technical literature show that most canned baby foods meet that standard and that the lead content of canned products, other things being equal, depends on the amount of solder exposed to the food and on the acidity of the product. The new welded cans have eliminated solder altogether and products packed in glass contain less than 0.1 mg of lead per kg.

It has long been known that canned foods pick up tin. Indeed, the characteristic flavour of canned asparagus is due to it. This pick-up and the unsightly staining caused by some products, notably fish, led to the introduction of internal can lacquers which reduced metal pick-up but introduced another possible source of contaminating compounds. Thus far, however, there has been little evidence of contamination from them provided the solvents are thoroughly dispersed. Current advances point clearly to the introduction of water-based lacquers but, whether water or solvent based, the lacquer itself is a plastic. Aluminium cans are, of course, with us in great numbers but aluminium itself is of no toxicological consequence and any lacquers used are subject to the constraints referred to below. Similarly, drums used for the storage and transport of raw materials and some finished products are lacquered, enamelled or lined with polythene. If the latter, the container, again, is in effect a plastic one.

Metal foils have been used with food for a long time. Pre-war tin foil was virtually the only one available but now it has been supplanted almost entirely by aluminium. Metal foils are usually coated with a suitable lacquer to protect the foil from corrosion by constituents of the food and to ensure a hermetic seal. There may be migration of plasticizers into the food so once more concern is more with plastics than with metals.

Cellulose: This group includes paper and regenerated cellulose more commonly referred to by the brand name 'cellophane'. Neither of these is of any great importance in terms of packaging migrants in spite of the possibility that sulphur dioxide may be picked up from certain types of paper. Reference has already been made to the concern over PCB contamination of paper (p. 53) and to the dangers of lead contamination from some printed papers (p. 48). Cellophane is in direct contact with some dry foods but is more usually a support for plastic films.

Glass and Ceramics: Glass is a very inert material but it is possible to dissolve tiny amounts of inorganic substances from it. They are negligible but the use of ultra-violet absorbers and silicone release agents as used on non-stick fry-pans needs to be kept under review. The appearance of chips of glass in food products is one of the production manager's fears, but such contamination is physical rather then chemical. The solution of heavy metals from glazes on ceramics has already been discussed (pp 48-51).

Plastics: The development of plastics pre-war, mainly the phenolformaldehyde resins, led to their introduction as packaging materials, but their use for food was very limited. The explosion of plastic technology post-war changed things considerably and suitable plastic containers and films were introduced for food, but until the late 1960s there was very little information anywhere about what happened to food packaged in them. By 1970 the British Plastics Federation had introduced a code of practice for safety in use of plastics for food applications and others were taking an interest. In general the attitude quite properly was that there should be no migration from plastics to food and if there were any suspicions that this did occur there should be an additive submission and limits would be set.

Within a year or two draft standards for polyethylene for dairy products were being suggested and interest in the possibility of stabilizers and plasticizers migrating was quickening. Then, in 1973, concern over vinyl chloride monomer in polyvinyl chloride erupted with the association of a rare form of cancer with the inhalation of VCM in PVC factories. There were reports of the transfer of residual VCM from PVC film and containers into food and drink and questions began to be asked about other possible migrants.

A great deal of work was done by industry and government in collaboration. This resulted in standards and regulations for VCM much as follows:

in PVC containers for food use not more than 10 mg/kg
in PVC films for food contact not more than 1 mg/kg
in food not more than 0.1 mg/kg (since reduced to 0.05)

Representatives of government, consumers, the food industry, the plastics industry and those responsible for standards as distinct from regulations all contributed. Since the food industry bears the responsibility for all that appears in the final product, it combined with the government to press the plastics industry, those who made the plastics and those who turned them into film and containers, to put their house in order.

Other plastics also have come under scrutiny, particularly the acrylonitrile-butadiene-styrene (ABS) polymer used freely in tubs for yoghurt, cream, margarine etc., and again there has been close collaboration to get levels of acrylonitrile down to recommended limits of not more than 10mg/kg in containers and not more than 0.02 in foods and beverages.

In all these discussions it is notable that information available to governments in other countries is freely available to all with the result that the regulations appearing round the world vary very little. Recent emphasis has been on the carcinogenic monomers, but plastics, as is well known, contain colourants, stabilizers, plasticizers and so on. Heavy metals from the pigments are covered by the normal regulations. Stabilizers and plasticizers are covered by a general regulation forbidding the presence in food of anything not specifically permitted or are specifically regulated but adherence to guidelines or standards for Plastics for Food Contact will ensure that all packaging materials delivered to food manufacturers will be within the regulations laid down and that the food packed in them also will comply with those regulations.

While Australian Standards have been issued for several polymers used in food packaging, new knowledge will always lead to change and as this book is being written there is concern over reported carcinogenicity of phthalates which are found in a number of plastic formulations. Consequently, these substances are now under review in a number of countries.

All this activity to control the migration of unpleasant substances from plastic containers into food can be nullified in the home by the consumer who uses for food purposes plastic receptacles manifestly not intended for food use. The manufacturer is under no obligation to consider the formulation of garden equipment, rubbish bin liners, etc. Do not, therefore, put food into plastic containers designed for other purposes. Don't mix that party punch or home brew in a plastic rubbish bin or use a laundry basin for the bulk preparation of fruit salad.

4 Food Additives

Feed me with food convenient for me.

Proverbs 30:8

Not so long ago a medical practitioner rang a food technologist and asked him what preservatives were used in his company's canned foods. This astonished the technologist because heat processed canned foods do not contain preservatives; they are not permitted to contain preservatives; and, in any case, to add preservatives to a product about to be preserved by heat processing would, at best, be silly. The incident did, however, illustrate the depth of ignorance in the community about food processing in general and about food additives in particular.

What are food additives? They have been described as 'any chemical substance that enters into a food'. This is an inadequate definition, as it would include many naturally occurring toxicants, such as scombroid and paralytic shellfish poisons, grass and other feed taints in milk and all chemical contaminants. The Joint Expert Committee on Food Additives (JECFA) defines additives as 'non-nutritional substances which are added intentionally to food', but sorbic acid, a preservative, supplies energy, and vitamins and minerals are nutrients in their own right. There is no short definition; it is better to expand it thus; food additives are those substances deliberately added to food by the manufacturer to facilitate processing or to improve appearance, texture, flavour, keeping quality or nutritional value. The rest are contaminants and there is little to be said in their favour though, as we have seen, there are groups of substances, such as boiler water additives, sanitizers and detergents, which are legitimately used, and even necessary, in food factories and may therefore find their way into foods. Because of this they are treated as additives and permission must be obtained for their use in food manufacture (see Chapter 7).

As we have already seen in Chapter 1, the use of food additives is not new. What *is* new is the proliferation of food additives in a much more highly sophisticated form as urbanization has increased, as the ability of individual families to feed themselves has disappeared, and as consumer demands have spurred on technology to more sophisticated, attractive and convenience foods.

It is clear that food additives have always been used functionally; to colour, to flavour, to thicken and to preserve. Colour, flavour, texture and keeping quality have always been important to the manufacturer and vendor alike. They remain of prime importance because they, and other properties conferred on food, relate to the consumer's expectations. To a very large extent, the basis of modern food additive usage is sociological, not technological, for the consumer wants the results which only food additives can guarantee.

Table 4.1 sets out the major classes of food additives with the particular property of food to which they contribute. Some, such as colours, flavours and preservatives, perform one function only, others, such as emulsifiers, aerators and acidulants, contribute to more than one property. While the reasons for the use of food additives are clearly implied in this table it is time to consider them in more detail.

Colours

Colour is a most important property of food. It has a great deal to do with the selection of food, as the contemplation of human behaviour at any smörgåsbord will confirm. More than that, however, tests on the manipulation of the colour of food by dyes and coloured lights have clearly demonstrated the great influence of colour on the acceptability of food and on the perception of flavour. There is reason to believe, also, that colour may be of physiological value in the stimulation of appetite.

The place of colour in food presentation was recognized long ago by cooks and chefs and its extension to the market place goes back a long way also. Its value to caterers, food manufacturers and others commercially interested in the sale of food is exploited every day. Colour is necessary; it is part of the pleasing appearance of food and it helps to sell products by eye appeal, but the aversion of the consumer to products which are over-coloured also is known.

But colours have been used to deceive. Medieval cooks simulated egg custard with the help of saffron and marigold and deception in commercial products was rife in the nineteenth century. Echoes remain today in the

Table 4.1: Food Additives and Their Functions*

	Appearance	*Texture*	*Keeping quality*	*Flavour*	*Nutritive value*
Acidulants alkalis and buffers			+	+	
Aerators	+	+			
Anti-caking & free-running agents	+	+			
Antioxidants			+		
Colours	+				
Emulsifiers gums & stabilizers	+	+			
Enzymes		+		+	
Flavours				+	
Humectants		+	+		
Minerals					+
Preservatives			+		
Sequestrants			+		
Sweeteners				+	
Vitamins					+

* (Expanded from K. T. H. Farrer, 'Food Additives – For and Against', *Food Technology in Australia,* vol. 27, 1975, pp 379-87)

permitted use of colour in custard powder but, nowadays, the presence of the colour must be declared on the label.

The basis for the use of food colour is essentially aesthetic and, strictly speaking, there is no technological justification for adding colour to food at all. However, tests have been carried out with food products which are normally coloured but from which the colours have been omitted and the dramatic falls in sales experienced have shown conclusively that the ordinary consumer has clear-cut expectations of what specific products should look like.

Given this then, it can be said that there are three technological arguments in favour of colour addition, as follows:

1. To intensify natural colours considered by manufacturer and consumer to be too weak, though this is an aspect of the aesthetic or 'eye-appeal' argument;
2. To overcome variations in colour intensity in the raw materials, i.e. for uniformity which is very important in the marketing of a food product.
3. To replace colour lost in processing or on storage by
   - the action of heat
   - bleaching by, say, sulphur dioxide used as a preservative
   - the action of light.

These are, however, weak technological arguments and food colour usage remains firmly based on the expectations of the consumer and the desire of the manufacturer to sell his products by meeting and even promoting those expectations. There is therefore a history of the colouring of food which issues today in regulations which permit some foods to be coloured. But with what?

As has already been noted, the first colours were mainly of vegetable origin. Then mineral colours, lead and arsenic salts especially, appeared and were used at first in ignorance then callously until analytical chemistry developed in the nineteenth century and put a stop to it. Finally, the synthetic dyes, the so-called coal tar dyes, appeared. The description, coal tar dyes, is still used almost as a term of abuse and certainly in a condemnatory way; it is therefore worth looking more closely at it.

Because it is dug out of the earth, coal may be regarded as a mineral and, indeed, is so defined by the *Shorter Oxford Dictionary*. But it is also a plant product and this, too, is recognized in its dictionary definition. It is true that its chemical properties have been changed by pressure, heat and time from those of plants, but it should be no surprise that a lump of coal contains substances with the potential for being reorganized under favourable conditions into plant products such as colours, flavours and vitamins.

When coal was first dry distilled to make coke for smelting, gas and tar were obtained and the latter, especially, was found to be a storehouse of interesting chemical substances from which others could be made. In 1856 William Perkin, seeking to make quinine, a plant product, stumbled on 'mauveine', the first synthetic dye. Others followed quickly and it was not long before indigo was made in the laboratory. This was a dye made from plants of the *Indigophera* species which were grown in plantations in India for dye production, but the synthesized dye was exactly the same as the natural colour and an industry disappeared. Similarly, thiamin (vitamin B_1), obtained initially from rice polishings, has long been synthesized from substances found in coal tar, and vitamin C, still derived in sufficient dietary quantities from fruits and vegetables, is also easily built up in the laboratory. These compounds, and others like them, could be described as coal tar products but the derogatory connotations of such a description seem to have been reserved for colours.

Chemical ingenuity led to the synthesis, from starting materials obtained from coal tar, of whole families of new dyes. Some were used in foods. Thus, at the turn of this century some eighty different textile dyes were being used in foods. No one had thought about any relationship between chemical structure and cancer and, in any case, because of their high tinctorial power, the concentrations used were very low. Rational selling by the manufacturers reduced this number but those remaining, and natural colours, too, for that matter, have had to be reassessed in recent years for use in food and many have been banned. These include a beautiful red colour, rhodamine B, which, because of its stability and shade, was a favourite with the food industry. However, new information in the 1960s threw doubt on its long-term safety and it had to go. Such re-examinations will continue as new criteria for assessment of suitability for use in food are developed, but it must be emphasized again that the criteria are chemical structure and experimental evidence, not the origins of the materials from which the substances have been synthesized or the false confidence engendered by compounds of natural origin.

The last word on 'coal tar' origins must surely be the identification of phenol (carbolic acid) and cresol, two of the major constituents of coal tar distillates, among the many compounds occurring naturally in tea.

Colours were among the first food additives to be questioned when in the 1950s government regulatory officers all round the world began to look hard and long at food additives. The fat-soluble 'coal tar' dyes were the first to go and the next few years saw a significant shake-out in the lists of colours permitted to be added to food. Because the information on which decisions were made was essentially the same in each country, the lists which emerged, even after further revision, bear a strong resemblance to each other. Interpretation of the information available plus specific national constraints, such as the Delaney Clause in the United States which has led to the very small list of synthetic colours available for use in America, provides some variation. The countries of the EEC show little variation and, for the same obvious reasons, the permitted lists in Australia and New Zealand are essentially the same.

Flavours

Flavour is more important than colour. Eye appeal will sell a product once, but if the flavour is unacceptable for any reason at all the product will not be used or bought again, and the most nutritious food in the world has no chance of being nutritious if people will not eat it.

The technological arguments for flavour are those already set down for colour: to intensify natural flavours, to ensure uniformity of flavour and to replace flavour lost during processing. To these must be added a fourth, the addition of flavour to an otherwise uninteresting formulated new food; meat analogues from spun vegetable protein is an example.

The importance of flavourings in the history of food has already been noted and those traditionally used were natural plant products. What is so often overlooked, however, is that those same natural flavouring materials are mixtures of hundreds of identified organic chemical compounds. Many of them were identified and synthesized in the nineteenth and early twentieth centuries before they were found in fruit, essential oils and other plant tissue, including flowers, leaves, roots and bark, which had been used as flavourings in food for hundreds of years. Others have been synthesized since. Together they are available for use in the flavours offered to the food industry and together they give rise to the charge, frequently distorted for sensational purposes, that some two thousand chemicals are being added to our food. Most of them have been there for centuries; it is only the advance of science which has enabled them to be identified, characterized chemically, synthesized and offered for use as separate entities.

It cannot be too highly stressed that such naturally occurring compounds built up in the laboratory are exactly the same substances as those found in nature; the body cannot differentiate between them. Assuming that there is no question of imbalance (see p. 100) the only thing at issue, and this is sometimes important, is whether or not the synthetic product carries any impurities. But the same techniques which have led to the recent great advances in flavour chemistry are available, and are used, to ensure the purity of the products used to flavour food.

A great proportion of food flavours is, however, made up of the natural products themselves; essential oils, for example. These oils, known since the thirteenth century, are prepared by the physical processes of distillation, steam or dry or under vacuum, or pressure, and are familiar and valuable; citrus oils from orange, lemon, lime and grapefruit, oil of clove, cinnamon and spearmint are examples of oils used extensively in the food industry.

Because of the complexity of flavours, both natural and formulated, and because of the very low concentrations necessary to provide a desired effect, their regulation is very difficult and, in an attempt to bring some rationality into the matter, the International Organization of the Flavour Industry (IOFI) based in Geneva has suggested a five-fold differentiation as follows:

1. Aromatic raw materials of vegetable and animal origin: spices, such

as pepper and cinnamon, and meat extract are examples. They may be used as they are or as raw materials for concentrates etc.

2. Natural flavours such as concentrates prepared from aromatic raw materials by physical means such as extraction and concentration. They are complex mixtures of hundreds of chemical compounds.
3. Natural flavouring substances which are isolated from aromatic raw materials by physical methods. They include the essential oils and they, too, are complex chemical mixtures.
4. Nature identical flavouring substances which are prepared by synthesis in the laboratory or isolated by chemical processes but are identical with the natural substances. Vanilla is the best example. Formerly prepared from the vanilla bean, it is now produced chemically from lignin, itself a plant substance.
5. Artificial flavouring substances which have not so far been identified in natural products but which simulate identifiable natural flavours.

Most of the flavours used in the food industry fall into the first four classes either absolutely or as blends, i.e. they are not artificial but are naturally occurring complex mixtures. Nevertheless, in Australia and New Zealand any food to which they are added must be labelled 'Artificially Flavoured'. This is confusing until it is pointed out that the declaration refers to the act of flavouring. It is this which is artificial, not necessarily the flavour used.

It is worth noting that flavours are sold and used in solution and that *solvents* permitted to be used are regulated. Alcohol, ethyl acetate and glycerine are some of them, but, as every cook who uses vanilla essence knows, the amount of solvent appearing in the dish that reaches the table is very small indeed.

Sweetening agents are a class of flavours which calls for separate treatment but mention must be made here of *flavour enhancers*. These are substances which have little flavour themselves but which have the property of intensifying or enhancing other flavours.

Although several new enhancers have been introduced in recent years, the most familiar and by far the most used is monosodium glutamate (MSG). Oriental cooks in antiquity used a seaweed to make a stock which added richness to food flavours and early this century a Japanese scientist found that this property was due to glutamate. Shortly after, MSG was produced commercially.

Glutamate is an important constituent of all proteins, is therefore consumed normally by everyone every day, and is abundant in the human body. It occurs free in mushrooms and tomatoes.

Some years ago individuals began to associate feelings of tightness, warmth and tingling with the use of MSG. Because restaurants serving oriental foods often supply shakers of MSG for use with such dishes the phenomenon was nicknamed the 'Chinese restaurant syndrome' and doubt was thrown on the safety of the substance. Subsequently, there was a stir when animal experiments suggested that MSG could affect the central nervous system. The amounts required, however, were massive and, after much work on acute and chronic toxicity, and a detailed search for other adverse effects, the conclusion is clear. At all concentrations used in foods, MSG has no adverse effects except on the very small proportion of people who are exquisitely sensitive to it.

Sweeteners

Sweetness is a flavour and sugar (i.e. sucrose from cane or beet), though it contributes energy, is a most important flavouring substance. However, it poses problems for the diabetic and the overweight and contributes to dental caries, so other sweetening agents are sought. Some, such as sorbitol, are sugar-like but less sweet than sugar, make lower insulin demands and are possibly less prone to cause tooth decay (i.e. are less cariogenic). There is a case for their use as food ingredients but diabetics and those with weight problems have long relied on non-nutritive sweeteners to replace the pleasant flavour of sugar which they would otherwise be denied.

A non-nutritive sweetener is a compound which tastes sweet but which is not itself a sugar or a polyhydric alcohol such as sorbitol. The most familiar is saccharin which was discovered in 1879 and commercialized in 1900. It is quite a simple molecule and, within the limits of detection, is excreted unchanged. It is 250 to 700 times the sweetness of sugar depending on the conditions of usage but leaves a slightly bitter aftertaste on the palate. It is used in diabetic and low-energy foods and is well known as a 'table top' sweetener. It is carried by many people who add it as tablets to tea or coffee.

The cyclamates, which were discovered by accident in the 1940s and introduced to the food industry in 1950, offer an alternative to saccharin. They also are relatively simple chemical compounds but only 24 to 60 times the sweetness of sugar. Cyclamates, however, have the property of masking bitterness and they rapidly became popular in low-energy foods, especially soft drinks in which saccharin was used less because of the bitter aftertaste. In 1968, however, it was reported in the United States that cyclamates could cause bladder cancer in animals and an extraordinary trial by news media followed. The Americans, caught by their own Delaney Clause (see Chapter 5), banned cyclamate from general use and, to the disgust of the prestigious English newspaper of science, *Nature*, even the United Kingdom authorities succumbed. In fact, the doses required to produce an effect in susceptible rats were enormous, and cyclamates continued to be permitted in more than thirty countries. In Australia and New Zealand the use of non-nutritive sweeteners has always been closely regulated and confined to specified foods, so, with all the scientific evidence available to them, the authorities in both countries made no change to local usage.

In 1971 it was saccharin's turn to be accused of producing bladder cancer in rats. Much work has since been done and a Canadian study suggested a correlation with bladder cancer but again the doses were huge. Again, however, the Americans were obliged by the Delaney Clause to ban the compound, but this time the public rebelled. There was no alternative for the diabetic and the weight-conscious and the United States Congress in 1977 forbade the Food and Drug Administration to ban saccharin so that administrative devices for postponing action had to be used. One somewhat exasperated United States senator is said to have proposed that all foods containing saccharin should bear the label 'The Canadians say that saccharin is bad for your rat'.

In Australia and New Zealand cool assessment of the scientific evidence led to the conclusion shared then in the United Kingdom and elsewhere and now by the American Council on Science and Health that, as used in food, there is no danger from saccharin at all.

In the meantime, another non-nutritive sweetener, aspartame, is gaining acceptance. Also discovered by accident, aspartame is related to the proteins and is thus chemically quite different from saccharin and the cyclamates. Several other compounds have been suggested and no doubt some eventually will be found to be acceptable, but many years of testing lie ahead of anyone wanting to introduce any additive.

No non-nutritive sweetener or blend of sweeteners can replace exactly the flavour of sugar, a natural food product, in food systems. Nor can they match the 'mouth feel' of sugar in beverages. They remain, however, valuable additives for the specific purposes already referred to.

Preservatives

The need and desire for the preservation of food to provide man with sustenance during the dark and frugal days of the northern winters are as old as man himself. Sun-drying, smoking, salting, pickling and fermentation (including cheese and wine making) are all of great antiquity. Chemicals have been used deliberately only quite recently and in only a small proportion of preserved foods (see Table 4.2).

The first preservative was almost certainly wood smoke, as prehistoric man found that food affected by the smoke of his fires kept marginally longer than fresh food. Very slowly, by trial and error, a deliberate

Table 4.2: Major Methods of Food Preservation*

Method	Effect
Heating	Killing of organisms
Refrigerating	Inhibition of growth
Drying Salting Syruping	Inhibition by denial of water
Acidifying (pickling)	Inhibition of growth
Chemical preservatives	Killing or inhibition

* (From K. T. H. Farrer, 'Is our food safe?' in *Food Quality in Australia,* Australian Academy of Science Report Number 22, 1977)

smoking process developed. Smoke houses were built and in them fish and meat were exposed to the vapours of smouldering wood and sawdust. These vapours contained hundreds of chemical compounds and the type of wood, moisture, air supply and temperature all influenced not only the production but also the composition of the smoke and hence the flavour, rind formation and keeping quality of the product.

Wood smoke repels insects, has an antioxidant effect and kills microbes. On these properties its preservative effect depends and that effect has been attributed mainly to the phenols present; but the chemistry of smoked foods is further complicated because the many chemicals in the smoke react with each other and with some of the normal constituents of foods. In this way they contribute not only to the keeping quality but also to the flavour and colour.

Smoking presents a complicated chemical picture involving a large number of undesirable substances. Many of them are poisonous at higher concentrations and would never be permitted as individual additives. Some are carcinogens, the most common of the latter being 3, 4-benzpyrene which is easily identified in smoked products up to concentrations of 50-60μg/kg.

Fortunately, perhaps, heat processing and refrigeration have reduced smoking to a minor preserving technique but it is still used because of the attractive flavour of smoked foods. There remain, however, the backyard barbecues and one shudders to think what might be absorbed by various meats from the smoky flames of many of them. It is far safer to use a hot plate over a bed of glowing embers.

Salting is another very old method of preservation. Thousands of years ago the Egyptians were preserving flesh with salt and salted fish was a common article of trade in the ancient world. The Romans had factories for making a highly salted fish sauce and paid an allowance to their soldiers for the purchase of salt. This was the *'salarium'*, hence salary. Christ called his followers 'the salt of the earth' because, though small in number, they could, by the quality of their living, influence the 'flavour' of the society in which they lived. In medieval times to be seated 'above the salt' at the long communal tables was a mark of esteem. And so on. Salt has an important place in many communities, but from ancient times salt made possible the preservation from season to season of meat and especially fish. Later, travellers, particularly ocean travellers, benefited. Australia and New Zealand could scarcely have been colonized without salt meat and for twenty-five years the trade in salt pork from Tahiti, initiated by Governor Hunter in 1801, sustained the colony of New South Wales.

Salt is a compound of sodium and chlorine and in recent years the medical profession, nutritionists and dietitians have become increasingly concerned about the physiological implications of sodium and, in particular, about its positive relationship in some people with hypertension. Both sodium and chlorine are essential in every diet but there is sound evidence for believing that some people are genetically disposed towards salt-induced hypertension and salt must therefore be closely controlled in their diets.

On the other hand, others seem to be able to consume as much salt as they like. The question is a medical one and must be settled for each individual by his or her own doctor. In the meantime, those who urge that salt should not be added to foods, especially processed foods, would do well to keep four things in mind.

1. Much of the salt added to food is added in the home; either in the kitchen or at the dining table.
2. The food manufacturer will not make products which he cannot sell and salt is an important flavour enhancer.
3. There are foods which cannot be made without salt.
4. Salt is still an important preservative.

The scientific basis for the preservative effect of salt is now well understood. Put simply, water is necessary for supporting the growth of organisms and if the water in the immediate neighbourhood of the organism is concentrated by drying out some of it, or freezing some of it, or by adding water soluble material such as salt (or sugar), the water available for the organism is reduced, i.e. the 'water activity' falls. For every organism there is a minimum water activity below which it will not grow. Ultimately, a value is reached below which no organisms at all will grow and the food is preserved. Salt is the most efficient substance available to the food technologist for this purpose but sugar also is used in jams, syrups and glacé fruits. Aerators (cream of tartar and bicarbonate of soda) in baked goods have the same effect, though this is not the primary reason for their use.

Some foods simply cannot be made without salt because salt is used so to control the microbial environment that only those organisms necessary to produce the particular food will grow. The many varieties of cheese, all of which are excellent sources of protein and calcium, are possibly the best examples. Cheddar contains, say, 1.4-1.8% salt, Swiss 0.4-1.3%, Camembert 2.4-2.7%, Blue 4.0-5.0%, Feta 4.5-5.5% and Romano 4.5-6.0%. Unless they contain salt at these particular levels the particular varieties named will not be obtained. The products will be something else. Salt is essential for the production of sauerkraut also and yeast extracts would never survive the constant dipping of knives without salt to protect them.

Most food manufacturers are well aware of the concern over salt and many formulae, especially those for baby foods, have been and are being modified as far as possible to reduce salt. But salt is so important to many flavour profiles that, in those foods which do not depend on its preservative effect, there is a limit below which it cannot be reduced without adversely affecting sales. Similarly, the replacement of sodium with potassium in some products has been achieved but is seriously limited by adverse flavour effects before the question of sodium/potassium balance in the body arises.

There are on the market many highly salted snack and party foods such as nuts, pretzels, potato chips etc. In them the salt is highly visible and their use is as controllable as the addition of salt in the kitchen or at the dining table. Should it really be necessary, the salt content of every processed food can be declared on the label but the responsibility for most of the salt consumed lies with the housewife.

The food preserver aims to kill micro-organisms or to stop them growing. Smoking and salting do both and so do chemical preservatives. Smoking preserves partly by the chemicals it transfers to the food and so did the pitch linings of barrels used for wine and beer. The resinated wine of Greece, retsina, derives from the preservative resins transferred to the wine from the wood of the barrels used and hops added to beer as a flavouring have a weak preservative effect. So do spices.

In the middle of the nineteenth century there was a great deal of activity on the chemical preservation of food and many patents were taken out. One man proposed to mix cheese thoroughly with rum, to immerse it in a bath of salt and saltpetre solution for ten minutes, to dry it, to cover it with boiled linseed oil and to wrap it in tin foil. Another used for preservative purposes a composition of white salt, saltpetre, white wine vinegar, acetic acid, rectified alcohol, pure white wine, Spanish pepper, black and white pepper, cayenne, paradise seed, cinnamon, ginger, calamint and pennyroyal. These at least could be regarded as condiments, but a third man used antiseptic agents such as tannic and gallic acids, oak bark, cinchona, mimosa bark, etc., either in the food of the animal before slaughter or by post-slaughter injection through the blood vessels.

The most favoured chemical at that time was sulphite and it has continued to be very useful. It is easy to use and it is effective in extending the shelf life of a number of products.

The term 'sulphite' has a precise chemical meaning but it is used loosely in relation to food preservation. When sulphur is burnt it forms sulphur dioxide, an acrid, irritating and corrosive gas which dissolves in water to form sulphurous acid. This acid forms a series of salts, for example with sodium hydroxide (caustic soda), which are called sulphites, in that case sodium sulphite. It also forms another series, the metabisulphites. These are 'richer' in sulphur dioxide which is very easily released from them into food and they are thus more effective as preservatives. It is the sulphur dioxide which is the preservative, it is simply most conveniently added as a sulphite.

Sulphiting was practised in ancient times by burning sulphur and allowing the sulphur dioxide to flow round the food to be preserved. Within living memory it was also burned in sick rooms to disinfect surfaces, bed clothes and patients.

In the middle of the last century many attempts were made to use sulphur dioxide and sulphites as preservatives in various foods, expecially meats. Mostly they failed because the true nature of food spoilage was not understood. Later, salicylic acid, boric acid and benzoic acid were used extensively, so extensively that in America a team of volunteers tried out their effects on the human body and in Australia concern over their use gave rise to the Victorian *Pure Food Act* of 1905 and its associated regulations, a world first. Salicylic and boric acids have long since disappeared from lists of permitted additives but sulphites and benzoic acid remain. They are partly interchangeable, but only partly because neither is effective against the full range of spoilage organisms.

Sulphites are particularly used in wine, fruit juices, sausage meat and sausages but are not permitted in minced meat. To some this seems to be an anomaly but sausage meat may contain starch in the form of flour as a binder which, with the rise in temperature caused by the fine mincing, makes the uncooked product particularly susceptible to microbial spoilage. Sulphite also plays a valuable part in preventing the darkening of peeled fruit and vegetables between peeling and processing. There is always some residual sulphite in products so treated and it must therefore be declared. But the sulphite is not there as a preservative at all; it is simply the residue of the sulphite used, in this case as a processing aid.

Sulphite is convenient and effective but its high level of use is under review because of reports that sulphite in foods may trigger asthma attacks and cause gut lesions in susceptible individuals.

As already noted benzoic acid, which occurs naturally in some fruits used for food, is partly interchangeable with sulphite. It is not used as much as sulphite but it has the advantage of being more permanent than the latter which slowly changes to the ineffective sulphate.

Though sulphite has particular relevance in sausage meat, the nitrates and nitrites are preferred for most meat products where preservatives are required. Sodium nitrate (Chile saltpetre) has been used for centuries in the curing of meat. Nitrate reduces readily to nitrite, which inhibits the growth of the deadly micro-organism, *Clostridium botulinum,* in cured meat products, improves the colour of the products by reacting with muscle myoglobin to give a characteristic pink colour, contributes to the flavour of cured meats and helps to control off-flavours by acting as a mild antioxidant. Though not required to the same extent to control botulism from canned meats, its use in these products has been permitted because of its beneficial effect on the colour and because it confers a measure of protection on the meat after the can has been opened.

In 1969 the American Meat Institute warned that nitrosamines may be formed in cured meats from a reaction between added nitrites and naturally occurring secondary amines. Because nitrosamines are known to be carcinogenic in rats this announcement caused some concern and, in some quarters, emotional opposition to the continuing use of nitrate and nitrite as food additives. The problem is that refrigeration plus nitrite and mild heat treatment plus nitrite have established an excellent public health record for cured meats and that the elimination of nitrite would, on the face of it, increase significantly the risk of botulism. In addition, bacon is the only frequent source of food-borne nitrosamines at any level and then only in micrograms per kilogram (or parts per thousand million).

It was soon found that nitrosamines are formed relatively easily by reaction in acid conditions such as occur in the stomach and by bacterial activity in neutral conditions. Whether or not the concentration of nitrosamines likely to be present in the gut from all sources is a hazard, anyway, is a moot point.

In 1978 the results of a study carried out suggested that nitrite itself might be a carcinogen and that, in any case, it interfered with the immune system of rats fed nitrite at a level of 250 mg/kg of the total diet. Naturally enough, these results are being thoroughly checked, especially as calculations within the same department predicted an average daily exposure to nitrite of 4.6 mg from an average diet, 5.5 from a diet high in cured meats, 21.4 from a vegetarian diet and 13.3 from a diet associated with a high nitrite water supply. Add to this the fact that nitrite derived from such dietary sources circulates continually in normal saliva and the unreality of the situation is complete. Certainly, the advice of the American Consumers' Union to cook bacon in a well ventilated room to avoid inhaling vapourized nitrosamines seems extreme, to say the least.

Alternatives to nitrite in the production of cured meats from *Clostridium botulinum* are actively being sought, so far without complete success though mixtures of nitrite and sorbate have their champions. What has been done is to reduce the allowable limits of both nitrate and nitrite in meat products. In a number of countries nitrate has been eliminated from most meat products and reduced limits have been placed on nitrite, e.g. 125 mg/kg (calculated as the sodium salt) in perishable canned cured meat and 50 mg/kg in commercially sterilized canned cured meat. Nitrate is still needed in dry cured products such as salami. These seem to be reasonable levels, especially in view of the American calculations of nitrite exposure.

Finally, a word or two about anti-mycotics, i.e. substances which prevent the growth of moulds. Of these sorbic acid is the compound of choice. It is closely related chemically to the sugars and like them is burnt in the body to form carbon dioxide and water with the release of energy. It is used in cheese wrappers to prevent mould growth on the surface of the cheese and in pickles and similar products for the same reason. Because it is slowly broken down by chemical and microbial decomposition its preserving effect is limited and, in any case, it is not effective against the full range of spoilage organisms.

Antioxidants

Antioxidation is a special case of preservation. Edible oils oxidize readily in air to produce unacceptable flavours (referred to generally as rancidity), unacceptable colours and sometimes unacceptable textures. Many chemical compounds, some natural such as vitamin C and the tocopherols, have the property of protecting oils from this oxidation. They are antioxidants. Sometimes two or more substances act together to produce an effect greater than the sum of the effects of the two acting separately. This is called *synergism* and the substances involved are *synergists*. These are general terms used to describe the reinforcement of any kind of activity but the phenomenon is of far greater importance with antioxidants than with any other class of food additives.

There are many substances with antioxidant properties but few are suitable for food use. In general those so used are the gallates, butylated hydroxyanisole (BHA), butylated hydroxytoluene (BHT), tert-butyl hydroquinone (TBHQ), natural phospholipids, such as lecithin, vitamin C derivatives and tocopherols. Not all of those are suitable for all foods.

Antioxidants are usually limited to specific products and they are used in mixtures to benefit from the synergistic effects and with food acids, such as citric and tartaric, to tie up and thus render ineffective traces of metals which promote rancidity (i.e. as sequestrants, see p. 84).

Antioxidants are used in oils and oily and fatty products as permitted in the food regulations. In these they are especially valuable but they are also effective in other foods such as some fruit and vegetable products in which they may be used provided, of course, their presence is declared by name or E number.

Modifying and Conditioning Agents

Food additive regulations everywhere make provision for disparate groups of additives under general headings such as Emulsifiers, Stabilizers and Others, Modifying Agents, Food Conditioners and so on. These groups include emulsifiers and surfactants, gums, stabilizers and thickeners, mineral salts (including acidulants, alkalis and buffers), food acids and aerators, sequestering agents, humectants, free-running agents, quick release agents, enzymes and the solvents used for flavouring materials.

This is a formidable list which is a happy hunting ground for those who are seeking sticks with which to beat the food industry. But, of course, no formulated food product contains them all, or even most of them, added to which a great many of these substances are themselves food products or constituents of foods – lecithin, esters of fatty acids, pectin, starches, lactic, citric and tartaric acids, and phosphates.

But why 'modify' or 'condition' food? What is meant by these terms? Neither is adequately defined in regulations so we are forced back to our definition of food additives and their functions as set out in Table 4.1. From this and what follows it is apparent that modifiers/conditioners between them contribute to all the functions of food additives except nutritive value. Their greatest contribution is, however, to appearance and texture. Both are important. Failure of an emulsion leading to separation of oil or fat and the caking of powders detracts from the eye appeal of a product. Mayonnaise which will not spoon or pour satisfactorily, bread or cake which crumbles too easily, and the appearance of crystals of ice, salt or phosphate in a number of products, are all examples of poor texture. All are controllable by additives.

Emulsifiers, Gums and Stabilizers: When the chef combines the oil slowly with the seasoned egg yolk to make mayonnaise, he relies on the stirring to produce droplets of oil, on the lecithin naturally present in the egg yolk to help in dispersing the oil and on the protein of the egg to stabilize the resulting emulsion. Lecithin helps to lower the surface tension between the water and the oil thus dispersing the one in the other. It is 'surface active', i.e. it is a surfactant.

There are many other substances with similar properties and some of them are very suitable for food systems: glycerides from fats, phosphates, citrates and alginates from seaweed are some of them. They are used in a wide variety of foods: baked goods to emulsify doughs and batters, ice-cream, processed cheese, meat products, salad dressings and so on. Gums of vegetable origin such as agar, pectin, Irish moss (carrageenan), gum tragacanth, gum guar etc. and stabilizers and thickeners such as gelatine, starches and modified starches are used to stabilize many food systems. Jellies, sauces, fruit drinks, yoghurt, ice-cream, salad dressings and toppings are some of the products in which gums, stabilizers and thickeners are used to produce and maintain a desired texture.

Many of these foods have been made traditionally with natural products and most of the additives in the permitted lists of emulsifiers, gums and stabilizers are of natural origin. A few are products of modern technology but they are based on natural products such as fat or starch. All have been subjected to toxicological examination and all have been required to demonstrate technological advantages.

Acidulants, Alkalis and Buffers: A most important property of all biological systems, including food systems, is pH. Put simply, pH is a mathematical expression of acidity or alkalinity in aqueous solutions. It ranges from 1 to 14. pH 7 is exactly neutral; all values less than 7 are acid, the degree of acidity increasing logarithmically as the pH falls, and all above 7 are alkaline, the alkalinity likewise increasing logarithmically as the pH rises.

Obviously, the addition of an acid product such as vinegar (4% acetic acid), to, say, a salad dressing will send the pH down and increase the sharpness of the flavour. Similarly, the addition of an alkaline material like baking soda (sodium bicarbonate) or weak caustic soda solution will send the pH up. There are, however, substances which have the property of soaking up, so to speak, the acidity or alkalinity of such compounds as acetic acid and caustic soda so that the pH of the system does not change or changes much less than one would expect. Such substances are called *buffers*.

It will be apparent that all three of these groups of materials are important in formulating foods. Acids have been added for centuries as vinegar, lemon juice and verjuice (the juice of unripe fruit, especially grape and apple). Bicarbonates, especially bicarbonate of soda, have long been used as mild alkalis and phosphates of different kinds are most useful buffers. The greater understanding of food systems and the ability to measure pH have, in the last fifty years, given food technologists the power to control pH by the careful selection, particularly, of acidulants.

This term, acidulant or food acid, is preferred in this context because we are concerned with the acetic acid of vinegar, the citric acid of lemon juice, the lactic acid of whey and similar food acids as well as acid salts such as cream of tartar and some phosphates. All of them are very weak and gentle compounds compared with the dangerous and powerful mineral acids of the chemical industry such as battery acid and the spirits of salts of the plumber.

Acidity and pH are important in food for flavour reasons, including the standardization of the flavour of varying raw materials, such as tomatoes, and particularly because high acid, i.e. low pH, is a protection against the growth of microbial contaminants and an aid in heat sterilization. Fruits are easier and safer to preserve at home than beans because the pH is lower, i.e. they are more acid.

The food acids permitted in most countries of the world include acetic, citric, fumaric, lactic, malic, tartaric, phosphoric and their salts. One or two other similar substances are used for special purposes. Sodium, potassium and calcium salts of these acids and bicarbonate of soda are available also as alkalis and buffers. In any system their use is limited by the pH resulting from it.

Some of these substances, especially the food acids and some phosphates, also act as *sequestrants,* i.e. they effectively remove traces of metals from food systems and thus prevent unwanted reactions. These reactions include rancid oxidation in oils and dairy products, loss of some vitamins, discolouration and the development of off-flavours in some processed fruits and vegetables, bad discolouration in fish and shellfish, and a number of defects in aerated and alcoholic beverages and in dairy products and canned meat. If the metals are 'locked up' these reactions do not occur and shelf life is extended.

Aerators, or Chemical Leavening Agents: Sour dough was the leavening of the ancient world, wild yeasts producing the carbon dioxide which caused the bread to rise. In the nineteenth century yeast began to be grown specifically for the purpose but the chemical production of carbon dioxide in baked goods dates from the introduction of self-raising flour in the mid-1840s. The gas is produced by the action of an acidulant, cream of tartar (potassium hydrogen tartrate) on bicarbonate of soda.

Today, some dozen or so acid salts, mainly phosphates of various kinds, are used in dough systems to give faster or slower release of gas, but the source of the carbon dioxide is still bicarbonate, mainly sodium, sometimes ammonium and, for those on sodium-free diets, potassium.

Humectants: This word is derived from the Latin *humectare,* to moisten or wet, and substances which are humectants are therefore used to prevent food products from drying, i.e. to keep them moist. They also improve keeping quality by making conditions more unfavourable for microbial growth. The most common humectants are glycerine and sorbitol. The latter is a first cousin of the sugars and is metabolized by the body as such. Humectants are used in confectionery especially to prevent sugar confectionery from becoming hard and unpalatable. In the home, glycerine is added to icings for just this purpose. Humectants are also used on desiccated coconut and in fruit cake. A non-food application is the use of glycerine to keep tobacco moist.

Anti-caking and Free-running Agents

Many powdered or granular foods are prone to caking which is not only unsightly but may seriously militate against the function for which they were made. Salt, icing sugar, milk powder and cheese powder are examples of both retail and wholesale products in which caking is both common and unwanted. Fortunately, substances are available to guard against this defect and to ensure the free flow of powdered product where it is wanted in the home and industry.

These substances are inorganic (i.e. mineral) compounds, mainly phosphates and silicates. They ensure that salt comes out of the salt shaker easily, that the various powders flow readily into the cups from tea and coffee bars and that feed stocks run cleanly to packaging and tableting machines in factories.

Enzymes

Enzymes belong to a special class of proteins produced by living cells and have the property of catalyzing specific biochemical reactions in plants and animals. They are therefore found in all raw materials used by the food industry and are responsible for a great deal of flavour development and many desirable texture characteristics. However, we are not concerned here with enzymes naturally present or even with those contributed by, say, the yeast in breadmaking but with those specifically added to food. The most familiar of these is the humble junket tablet containing rennin, or, to give it its modern name, chymosin.

Chymosin is the active principle of the cheesemaker's rennet which is obtained from the stomachs of calves and without which cheese would never have been made. It is still the most important enzyme used to produce cheese curd from milk but in recent years enzymes of vegetable origin, especially from specifically grown moulds, have been successfully used.

The amylases are a group of enzymes which act on starches. They may be prepared from plants, bacteria or moulds and are used in the manufacture of dextrose and sugar syrups, in baking, in brewing and distilling to reduce the starch to fermentable sugars, and in confectionery. Invertase is used to make glucose from cane sugar and lactase is used for splitting the lactose of cheese whey into a sweeter mixture of glucose and galactose. This enzyme has become important as advances are being made in the utilization of whey from the cheese industry.

Pectic enzymes which digest pectin and similar substances in fruits are used in the manufacture of a number of fruit products, including fruit juices, jams and jellies. Lipases act on fats and are responsible for much of the flavour of cheese. Some have been prepared commercially from animal sources and are added to cheese to promote the development of particular cheese flavours, especially those associated with Italian-type cheeses.

The other major group of enzymes is the proteases, i.e. those which act on proteins. They are used to tenderize meat, to modify the gluten of bread doughs, in brewing to reduce haze in beer and, with amylases, in the making of some Asian foods such as soy sauce.

All enzymes used are natural products. This does not make them safe and they are subject to the same scrutiny as other additives. All enzymes are, however, inactivated by heat so that heat processed foods are unlikely to contain any, whether those naturally present or those used in some preliminary process.

Vitamins and Minerals

Because of a tendency for some manufacturers to add vitamins and minerals to certain foods in unnatural proportions and then to make advertising claims based on them, regulations have been introduced in some countries to limit those vitamins and minerals which may be added to foods, to restrict the list of foods to which they may be added, to lay down the maximum concentrations of each vitamin and mineral which may be so added and to stipulate the exact form of any labelling claim which may then be made.

The vitamins which are permitted are A,B_1 (thiamine), B_2 (riboflavin), B_{12} (cyanocobalamin), C, D and niacin; the minerals are calcium, iodine, iron and phosphorus. The usage of both groups in foods is restricted to perceived nutritional requirements and the claims which may be made are limited to a specific form of words.

Both vitamins and minerals under these conditions are food additives. They are added to replace those lost in the preparation of raw materials or in processing (e.g. B vitamins in cereals, C in fruit juices), to standardize a product normally a vitamin source (e.g. B vitamins in yeast extracts), to bring the levels in a substitute up to those in a natural product (e.g. A and D in margarine) or to supply a perceived deficiency (e.g. iodine in bread).

There are two other inorganic additives which should be mentioned, chlorine and fluorine. Chlorine is added to water to reduce microbial contamination. Very low concentrations are used and the formation in the water of trihalomethanes and their significance have already been mentioned (p.60). Fluorine, however, is added as a nutrient.

Fluorine: It is arguable that there has been more sound and fury over the fluoridation of water supplies than over any other single food additive. Official inquiries have been held in several countries, one of the best being that in 1979-80 in the Australian State of Victoria, the *Report* of which, published in 1980, went into the matter very thoroughly indeed. It points out that man has always been exposed to fluorine in his diet, body fluids, tissues and skeleton, that fluorine is an essential trace element in the formation of teeth and bone and that fluorine in the diet is a major factor in reducing the incidence of dental caries. It confirms that the best effects are obtained when fluorine intake begins early in life and agrees that 'fluoridation of water supplies is the safest, most economical and effective way of distributing fluoride to the community'.

Finally, after thorough investigation of the evidence, the *Report* confirms that under the conditions of fluoridation of water supplies, i.e. at a concentration of 1 p.p.m. (1 mg/l) there is no evidence whatever of any harmful effect.

It is hard to understand why there has been such opposition to the fluoridation of water supplies when there has been no let or hindrance to the addition over several decades of iodine, the first cousin of fluorine, to bread and salt to protect people from goitre in iodine-deficient areas.

Miscellaneous Additives

This account of food additives has covered the major groups and most of the minor ones. Mention should, however, be made of some other categories, although the actual compounds used may be included in classifications already discussed. Thus, *firming agents* used with some canned products, notably tomatoes and pickles, include calcium chloride and alum, both minerals. Similarly, the *dough strengtheners* include bromates and iodates, also inorganic substances, and ascorbic acid (vitamin C).

Bleaching agents used to whiten flour include ozone and oxides of nitrogen formed by electric discharges, chlorine, chlorine dioxide and benzoyl peroxide. If the last of these is used, traces of benzoic acid (a preservative) are found in the flour. In some countries peroxides are used to bleach milk used in the manufacture of certain mould-ripened cheeses.

Lubricants, propellants and release agents could arguably be included with those substances absorbed during processing (see pp 60-2) but they are so intimately associated with the food as to be considered legitimately as additives. *Lubricants* include medicinal paraffin oil on machinery, and talc – a mineral – in certain types of confectionery. *Propellants* used in aerosols from pressure packs include freons and nitrous oxide. Both disperse very rapidly and amounts consumed from foods presented in these highly specialized and expensive packs would be extremely low. *Release agents* used on such utensils as bread tins are silicones.

It will have become apparent throughout this chapter that much, indeed most, food additive usage is related to the extension of the shelf life and the overall performance of food products. Taken together, the additives on any list look formidable but it must be remembered that only a few, perhaps only one, are used in a given product at any one time and that many foods contain none at all. Even so, the protestations of the food technologist that his use of additives is safe must be examined. This is done in the next chapter. Before leaving the additives themselves, however, some mention should be made of the irradiation of food. Unfortunately, consideration of this process, which is likely to be of increasing interest in the future, was clouded by its early confusion with food additives.

Food Irradiation

Soon after the end of World War II peaceful uses for atomic energy were sought. One of those proposed was the preservation of food. Projects were set up in a number of countries, including Britain and the United States, and that mounted by the United States Quartermaster Corps was probably the largest and most sustained.

It was quickly found that the smaller the organism the greater the input of energy required to kill it. Insects require more energy than mammals, insect eggs still more and the energy requirement to kill micro-organisms such as those which cause spoilage in food is highest of all. Food was subjected to atomic radiation from various sources and to X-rays and electron beams, i.e. it was irradiated, but although the organisms were killed and the food was therefore preserved, unpleasant and unacceptable flavours were produced. This was clear evidence of chemical change. While these first attempts gave food products which were clearly unacceptable on aesthetic grounds, the technology improved and effective results were obtained with irradiation of lower intensity. But the nagging question remained. Are there any changes of public health significance particularly in those products which would not be rejected because of poor flavour?

In 1958 the United States Congress in amendments to the *Food Drug and Cosmetics Act* misguidedly decided that irradiation adds something to food and therefore that food irradiation had to be regarded as an additive and treated as such. It has taken about twenty years to get back to the point that food irradiation is a process, not an additive, but at least it has become clear that the energy levels required in that process are well below those required to make food radioactive and that the short answer to the question posed above is, No, not at all. In fact, food irradiation has been scrutinized far more intensively than any other food process. Toast is made by irradiation with red and infra-red rays and the chemical changes which occur are far more significant than any found in food irradiated with ionizing radiations, but toast and toasting are so familiar as to be unnoticed. Ionizing radiations are unfamiliar and therefore suspect, especially in the natural human revulsion to the first catastrophic application of atomic energy, and their effects on food have therefore been studied in far more detail than those of, say, heat.

Exactly the same principles apply to foods preserved by irradiation as to those preserved by other means. They must be nutritious, safe-in-use and micro-biologically stable. Changes in nutritional value have been found to be similar to those encountered in other processes and on storage, and the same precautions concerning pre- and post-processing contamination must be observed with irradiated foods as with foods processed by other means. Irradiation, no more than any other process, may be used successfully to 'clean up' poor raw material. Poor raw material will always give a poor product, no matter what the process, and bacterial toxins in the raw material will remain in the product just as heat-stable toxins will remain after pasteurizing or canning.

The process as it is practised today uses either gamma rays from isotopes of cobalt (^{60}Co) and caesium (^{137}Cs), which differ in wave-length and frequency from sunlight and ultra-violet, infra-red and X-rays, or beams of electrons from accelerators. Both isotopes and accelerator are completely isolated from the food, no contamination is possible, and from both sources the substance being irradiated absorbs energy. The dose absorbed when 1 kg of matter absorbs 0.01 joule of energy is called a *rad*. This unit has recently been superseded by the *Gray* (Gy) which equals 100 rads, i.e. it is the dose absorbed when 1 kg absorbs 1 joule.

Low doses of irradiation (up to about 1 kGy) inhibit sprouting in tubers and root vegetables and the over-ripening of fruits, and kill insects. Medium doses (1-10kGy) reduce the total microbial contamination and the number of non-sporing pathogenic organisms, and improve the technological properties of foods. High doses (10-50kGy) are required for commercial 'sterilization' and the destruction of viruses. Some specific applications are shown in Table 4.3. It is clear from this table that control

Table 4.3: Radiation Doses for Specific Food Applications*

Food	*Purpose*	*Maximum dose in kGy*
Chicken	To prolong shelf life and to reduce numbers of pathogens	7
Cocoa	To reduce insect infestation	1
	To reduce the microbial load of fermented cocoa beans	5
Dates	To reduce insect infestation	1
Fish	To reduce total microbial load and certain pathogens	2.2
	To control insect infestation of dried fish	1
Mangoes	To control insect infestation, to improve keeping quality by delaying ripening, to reduce microbial load by combining irradiation and heat	1
Onion	To inhibit sprouting during storage	0.15
Papaya	To control insect infestation and delay ripening	1
Potatoes	To inhibit sprouting during storage	0.15
Pulses	To control insect infestation	1
Rice	To control insect infestation	1
Spices and condiments	To control insect infestation	1
	To reduce the microbial load and the number of pathogens	10
Strawberries	To prolong storage life by partial elimination of spoilage organisms	3
Wheat	To control insect infestation	1

* (Prepared from the Codex Alimentarius Commission's recommended International General Standard for Irradiated Foods, 1979, and the *Report of a Joint FAO/IAEA/WHO Expert Committee on the Wholesomeness of Irradiated Food, WHO Technical Report Ser. No. 659,* 1981)

of sprouting requires a dose of only 0.15 kGy, that insects may be controlled by a dose of 1 kGy and that significant reductions of microbial contamination may be effected with doses of up to 3 kGy. The higher doses required for the greater control of micro-organisms are also clearly evident.

The advantages of irradiation are that it does not heat the food, it has high penetration, its energy requirements are low and it does not leave chemical residues. It is not surprising, then, that since 1976 more and more countries have begun to give approval for the use of it for one food purpose and another. By 1981 twenty-one countries were permitting its use with at least one product. At that time the Netherlands was leading the world in permitted applications with more than twenty and the most common application (nineteen countries) was the prevention of sprouting in stored potatoes.

In 1980 the Joint Expert Committee on the Wholesomeness of Irradiated Food confirmed that changes in irradiated foods were small, similar in different foods and predictable and that there were no toxicological grounds for imagining that such products would in any way be a hazard to a consumer. This committee had been set up jointly by the Food and Agriculture Organization, the International Atomic Energy Agency and the World Health Organization specifically to study the question, and its findings are authoritative. Adoption of this process will, however, continue to be slow because of the emotional and irrational reaction of large sections of the public to anything 'irradiated' or 'nuclear'. This has been clearly demonstrated in market tests in the Netherlands and, while such resistance prevails, both public health authorities and the food industry will continue to be very cautious. But the safety of food treated by this process under the guidelines laid down by the Codex Alimentarius Commission of FAO/WHO is firmly established and irradiation will be used more and more as its specific benefits are more widely perceived.

5 **How Safe?**

> All substances are poisons;
> there is none that is not a poison;
> the right dose differentiates a poison and a remedy.
>
> Paracelsus

Safety is an emotive issue. One is becoming inured to the demands of people who want to ban something – be it nuclear energy, 2,4,5-T or a food additive, until it is shown to be absolutely safe. Yet these same people drive cars, fly in aircraft and cross bridges – none of which is absolutely safe. It is simply not possible to prove a negative and it is manifestly impossible to prove that something is not harmful – to someone, somewhere, at some time, under some conditions. This constraint makes it very difficult to counter assertions that such and such a substance is poisonous, causes cancer or is simply harmful to health. Such assertions may be, and often are, made without any evidence whatever. They cannot be *proved* wrong even when they *are* wrong. This chapter faces the question of safety and seeks to show how the safety of food may be assured even in the absence of proof of *absolute* safety.

The Perception of Safety

Because everyone eats, the safety of food is an especially emotive issue and questions of the safety of foods containing natural toxicants, chemical contaminants and food additives are, as we have seen, valid ones. It is well, however, to put them in perspective.

Table 5.1: Actuarial Assessment of Number of Deaths per Annum in United States from Causes Shown*

Cause	Deaths
Smoking	150000
Alcoholic beverages	100000
Motor vehicles	50000
Hand guns	17000
Electric power	14000
Motor cycles	3000
Swimming	3000
Surgery	2800
X-rays	2300
Railways	1950
General aviation	1300
Large construction	1000
Bicycles	1000
Hunting	800
Home appliances	200
Fire fighting	195
Police work	160
Contraceptives	150
Commercial aviation	130
Nuclear power	100
Mountain climbing	30
Power mowers	24
School and college football	23
Skiing	18
Vaccinations	10
Food colouring	0
Food preservatives	0
Pesticides	0
Prescription antibiotics	0
Spray cans	0

* (From Arthur C. Upton, 'The Biological Effects of Low-Level Ionizing Radiation', *Scientific American*, February 1982)

Table 5.1 ranks a number of sources of risk according to their actuarially determined contributions to the number of deaths annually in the United States. In a study published in an article by Arthur C. Upton in *Scientific American* of February 1982, these risks were assessed by three different groups of people and their perceptions of the four food-related risks are

Table 5.2: Perceived Ranking of Food-related Risks by Selected Groups Shown in Table 5.1*

	League of women voters	*College students*	*Business & professional club members*
Alcoholic beverages	6th	7th	5th
Pesticides	9th	4th	15th
Food preservatives	25th	12th	28th
Food colouring	26th	20th	30th

* (Prepared from data in Arthur C. Upton, 'The Biological Effects of Low-Level Ionizing Radiation', *Scientific American,* February 1982)

shown in Table 5.2. All three correctly placed alcoholic beverages high on the list, but not as high as they really are. All three gave much more weight to pesticides than they warrant and the students were far more fearful of food additives than their wiser elders. In fact, actuarial analysis attributed no deaths at all to pesticides, food preservatives or food colours.

If attention is confined to the food supply, the ordinary consumer will almost certainly name food additives as the prime hazard. In fact, they come at the bottom of the list. Table 5.3 sets out in descending order of risk food hazards as perceived by two sets of authorities. The first was the United States Food and Drug Administration in the person of Dr A. M. Schmidt, the Commissioner, who spoke for his organization at a symposium in London in 1975. The second was the Marabou Symposium

Table 5.3: Food Hazards in Descending Order of Risk

United States Food & Drug Administration	*Marabou Symposium on Food and Cancer, 1978*
Food-borne infections	Microbial contaminants
Malnutrition	Nutritional imbalance
Environmental contaminants	Environmental contaminants
Naturally occurring toxicants	Pesticide residues and food additives
Pesticide residues	
Food additives	

on Food and Cancer held in Sweden in 1978 as represented by Professor A. S. Truswell and others who assembled the conclusions. The latter attempted some kind of quantification, rating environmental contaminants a thousand times less risky than microbial contaminants or nutritional imbalance, and pesticide residues and food additives one hundred times less risky again. That is, risks from food additives are one hundred thousand times less than those from the micro-organisms which may contaminate food and lead to food poisoning (see Chapter 6).

It has been said that 'nutrient intakes form a continuum from lethal deficiencies to lethal excesses' (Hathcock). The concept of *Acceptable Daily Intake* (ADI), to which we shall return, is therefore based on the understanding that all chemicals are toxic but that their toxicity varies markedly. As has also been said many times there are no such things as toxic substances – only toxic concentrations. This raises the question of dose/response which was referred to briefly at the beginning of this book.

The human body has a remarkable facility for detoxifying chemical molecules, provided it has the time. Everything consumed thus stimulates a response as the body begins to metabolize it but the fact that so many substances stimulate no observable response in the quantities or concentrations ingested allows the toxicologist to establish a 'no-effect level'. This is the concentration in the diet expressed as mg/kg of body weight which may be consumed over several generations without producing any discernible effect. The other extreme is the LD_{50}, i.e. the dose which causes the death of 50 per cent of the group of test animals (p. 101). This is a measure of acute toxicity and is one of the pieces of information sought by toxicologists in making a judgement on an additive and establishing a limit for a contaminant. Other data, referred to later, are combined with it and, applying a safety factor of, usually, 100, are used to establish an Acceptable Daily Intake (ADI) which is defined as 'the amount of a chemical which may be ingested daily, even over a lifetime, without appreciable risk to the consumer in the light of all the information available at the time of evaluation. *Without appreciable risk* is taken to mean the practical certainty that injury will not result after a life-time exposure' (Vettorazzi).

'Without appreciable risk' means that there is some risk. This follows from the very nature of biological testing. Because of the impossibility, already referred to, of guaranteeing that *no* ill effect will ever be suffered by any individual anywhere at any time (from anything, by the way, not just food additives and contaminants), the establishment of a no-effect level is a matter of probability, not certainty. An apparent no-effect with a group of one hundred test animals actually means that there is a 5 per cent chance that three out of the hundred may have been affected. If a thousand animals were used, and no ill effects were observed, there was actually a 5 per cent chance that three out of the thousand may have been affected. The perception of safety, therefore, became the perception of risk, and the relative statistical probabilities of low risk under different experimental conditions are one reason why those charged with the evaluation of safety must know the experimental details and laboratory conditions under which given results are obtained (p. 101).

'Without appreciable risk' also raises the question of risk/benefit. Everything we do, whether we know it or not, is based on a risk/benefit assessment. There is risk in getting sunburnt; there is a risk in eating barbecued meat; there is a risk in crossing the road; there is a risk in any form of transport; there is a risk in getting out of bed and there is risk in staying in bed. But each risk is accompanied by a benefit. Many people accept the risks of sunburn because they want the benefit of a sun-tanned healthy appearance. Others gladly accept the risk in flying because the benefit of being able to travel a long way in the morning to return in the evening of the same day having completed their business or attended a function far outweighs the risk of death or injury.

So also in the food supply; the benefits of increased yields of crops through the use of pesticides far outweigh the risk to the ultimate consumer from the residues permitted to remain in the food, and the perceived benefits of having a number of products on the market throughout the year in the forms desired outweigh many times the very low risks stemming from the use of food additives.

There are at least two ways of classifying risks and behind these perceptions of risk lie the fears many people have about a number of things. First, there are the familiar and unfamiliar. What we may call the ordinary kitchen risks of scalding, burning, cutting oneself with knife, glass or tin, or being electrocuted by a faulty appliance, are familiar. Risks associated with nuclear energy are unfamiliar. So, one may say, are those associated with food additives, though they have been used for generations and the risks were once much higher than they are now. Secondly, risks may be voluntary or involuntary. The high risk associated with smoking is voluntary for the individual but involuntary for the non-smoker forced to tolerate a smoky atmosphere in, say, an aircraft. The risk attached to a food additive is voluntary only if the labelling of a product is complete, but no one labels for a contaminant and risks associated with them are involuntary.

Of the perceived benefits of food additives which have been referred to in Chapter 4, preservatives, stabilizers and convenience rank high with consumers, and flavours and colours are supported not only by marketing statistics but in the statement attributed to Hippocrates, 'An article of food or drink which is slightly worse but more palatable is to be preferred to such as are better but less palatable'.

It must be said, however, that the direct measurement of risk and benefit which the scientist would dearly like is not possible. There is no mathematical equation linking one with the other. The assessment of both is subjective, but not capriciously subjective. It is based for each additive and contaminant on considered scientific evaluation based on much experimental work and the combined judgement of many people qualified to make such judgements and moderating each other in arriving at a consensus. As we have seen, this consensus can never guarantee absolute safety, only 'safety-in-use', i.e., that there is a very high probability that this thing will be safe if used in this way.

The Evaluation of Safety

Prehistoric man must have made decisions by trial and error on which foods were safe to eat, but it is likely that someone at some time used animal behaviour and food selection as guides in the choice of foods for the family's use. We still use animals, but under carefully controlled conditions, to help us make decisions about substances in foods. Although it is anything but easy to apply to humans results obtained with animals, we are forced to do this because it is generally considered to be unethical to experiment on humans. There are exceptions to this and volunteers who are fully *au fait* with the implications of what they are doing sometimes take part in specific tests.

A famous example was Dr Wiley's 'poison squad' at the beginning of this century. Dr H. W. Wiley, who was not a medical man but an agricultural chemist in the United States Department of Agriculture, was caught up in the arguments about preservatives which raged at the turn of the century. In order to demonstrate the adverse effects of some preservatives and to support his efforts to secure the passage of a Federal law to control food, Wiley enlisted the help of a group of young men who consumed low doses of selected preservatives over a period of time. In due course they exhibited adverse symptoms which proved the point.

It was a drastic experiment which cannot be repeated today, especially because of the very large number of substances which are constantly under review.

General Considerations: A key role in the evaluation of food additives is played by the Joint FAO/WHO Expert Committee on Food Additives (JECFA). This committee was set up in 1956 to convene expert committees as required to consider problems associated with additives in food. JECFA is an international expert committee which, because of its composition, has access to a vast amount of information on the toxicology of many additives and the progress of toxicology and, because of its influence on regulation-making throughout the world, has made available to it even more information on the developments of food technology and on substances proposed for use as food additives.

Because toxicological studies depend on the care of colonies of animals and close attention over a long period of time to feeding régimes and regular examination, it is easy for work to become slovenly. Furthermore, unless an experiment is properly designed and the results carefully analyzed by means of the techniques of mathematical statistics, wrong answers may be obtained. Therefore, in assessing the significance of the work on a given additive, JECFA takes into account the quality of the laboratories, of the people doing the work and of the experimental design. Defective work lies behind the disappearance from the headlines of more than one ostensibly sensational finding.

JECFA has always sought as much information about an additive as possible. At its 25th meeting in Geneva in March 1981 the data required were systematized under the following headings:

Description
Raw materials
Method of manufacture
Impurities
Functional uses
Estimate of daily intake
Reactions and fate in food
Effects on nutrients
Substitute additives

The source of the information under each heading is also required and, in addition, specific toxicological and pharmacological information must be obtained.

Toxicology and pharmacology are relatively new sciences. They are related to the effects of poisons and useful drugs, respectively, on the body. Both sciences are inexact if only because of the wide variation of individual responses to doses administered and because of the difficulties of transferring to humans the results obtained with animals. Both sciences have, however, provided much vital information not only in the development of valued medicaments but also in the assessment of the relative hazards of contamination and of the safety of additives in foods. Day after day toxicologists are making successful predictions about hazards and eliminate from further consideration many otherwise potentially valuable substances.

Toxicology, bearing in mind our acceptance in Chapter 2 of the connotation of poisons as fast-acting compounds, is perhaps more concerned with the effect of a substance on the body whereas pharmacology, with its greater emphasis on the development of beneficial drugs, is more concerned with the body's response to a substance. Both, however, follow parallel and sometimes the same paths and both use similar methods. Thus, the first question of any substance under study is, what is its acute toxicity? But this is only one of a number of questions asked of any substance proposed as a food additive. Only if satisfactory answers are given to a series of such questions will the substance be considered.

The toxicological and pharmacological information sought by JECFA and used generally for the evaluation of the safety of food additives has been clearly set out by Dr G. Vettorazzi of the Food Additives (Food Safety) Unit, WHO, Geneva. It may be summarized as follows:

Acute toxicity. Toxicologists describe this as the LD_{50} which, as already stated, is the dose of the test substance required to kill 50 per cent of the animals on which the test is being carried out. Single dosings in several species of animals are required and the LD_{50} is usually described as the weight of the test substance per kg of body weight of the animal tested. Quite obviously, the higher the LD_{50} the less poisonous the substance.

Short-term studies. These studies are usually carried out on an animal from weaning to sexual maturity, say, 3 months in rodents and 1-2 years in dogs and monkeys.

Long-term studies. These span several generations of animals, say, 80 weeks for mice, 2 years for rats.

Biochemical studies. How is the substance absorbed by the body? How is it excreted – urine, faeces, bile? What is its half-life, i.e. how long does it take for half the quantity administered to be metabolized? What effect does it have on the enzymes which initiate and regulate so much of the body's biochemistry?

Special studies. Does it produce cancer? Does it cause fundamental changes in cells, i.e. is it mutagenic? Is it a nerve poison? Does it cause behavioural changes? Does it have any effect on any aspect of reproduction? Will it cause deformities in offspring, i.e. is it teratogenic? What is known about possible interactions with normal food constituents or other additives? Does it trigger or amplify any adverse effect caused by some other agent? Is there any significance in the change of balance between naturally occurring substances when extracted or compounded flavours are used in the formation of food products?

All these are very searching questions, and answers obtained from at least two species of animals and which will survive the scrutiny of critical scientists must be provided and added to any observations on man obtained by occupational or accidental exposure to such things as mercury, vinyl chloride monomer (VCM) and polybrominated biphenyl

(PBB), or from volunteers under special conditions. But this is not all. Because almost all substances are chemically altered or are broken down into pieces (degraded) in the body, these compounds, known as metabolites, must also be tested, as must any specific impurities. The work involved is therefore considerable and very expensive.

The methods used are to administer the substance under test in the animals' food or water, measuring the amounts consumed and hence the dose; or to deliver the test substance directly into the stomach through a tube, a much more certain way of ensuring that the dose desired is actually consumed; or to inject the substance in solution into veins, muscles or specific organs. When one wants information on a food additive it is obviously more satisfactory to see that it passes through the animal's alimentary tract. Injection raises the question of where and by what mechanism will the body absorb, metabolize and excrete the additive. The problems of translating results on animals into information of value in dealing with humans are big enough without complicating the issue by using an artificial method of administration. After all, milk injected into the human bloodstream could be expected to produce dire results.

A fundamental requirement of every test is that it be carried out on two equivalent groups of test animals. Both groups are treated in exactly the same way but only one group receives the test substance. The other is the control. At the end of the test the animals are killed and the organs minutely examined for abnormalities. The results are then treated statistically to ensure that differences between test and control groups are, in fact, of real significance.

The size of the doses used in these tests and the times over which they are administered are important, particularly in relation to tests for cancer production, but graded doses are used to establish both LD_{50} and no-effect level.

Having established a no-effect level for an additive, most regulatory authorities apply a safety factor of a hundred and permit in food concentrations of the additive only one hundredth of the no-effect level established in animals. This safety factor is sometimes less, sometimes more, than 100 but the concentration in food arrived at by its application is a regulatory limit, not a safety limit.

From all the considerations referred to above JECFA determines the Acceptable Daily Intake (ADI) which is allocated only when positive and sufficient information is available. When the data are insufficient a 'Temporary ADI' may be assigned or there may be 'No ADI allocated' depending on the nature of the insufficiency. Substances of very low toxicity may receive the designation 'ADI not Specified' but where the data available give no real indication of safety the term 'No ADI allocated' is used. 'Not to be used' is applied when there is sufficient clear evidence for supporting what is in effect a ban, and 'Decision postponed' is a designation which requires no explanation.

These classifications are thus judgements which are made by experts, almost all of whom are drawn from government laboratories or universities. Such classifications are under constant review as more knowledge comes to light. For example, at its 25th meeting in 1981 JECFA allocated a temporary ADI for the colour ammonium carmine (from cochineal) of 0-25 mg/kg. It was based on a no-effect level of 500 mg/kg obtained in a multi-generation study, but the very conservative safety factor of 200 was used because results for a long-term study were not available.

The LD_{50} is a measure of acute toxicity and a low LD_{50} is sufficient of itself to eliminate an additive from further consideration. It would not be proposed in the first place. No-effect levels and ADIs are based on the total corpus of information obtained from the required studies referred to above. Some of these must be considered separately.

Carcinogenicity: This is the capacity for causing cancer and compounds possessing such a property are called carcinogens. The most emotive of all the statements about food additives is that they may cause cancer. Some people say they do. Largely because of this fear, which is quite a recent one, the United States Congress in 1958 amended the *Food Drug and Cosmetic Act* by inserting what has become known as the Delaney Clause. This measure bans from food any additive 'found to induce cancer when ingested by man or animals, or if it is found after tests which are appropriate for the evaluation of the safety of food additives, to induce cancer in man or animals'. On the face of it, this is an inverted motherhood statement. Anyone for cancer? would be greeted anywhere with a thunderous and unanimous, NO. Unfortunately the Delaney Clause has become a millstone round the neck of the United States Food and Drug Administration and the ripples caused by the recurring convulsions in America have upset individuals and bedevilled regulation-making in many countries.

The Delaney Clause takes no account of which species of animals, how often they were dosed and under what conditions, or whether or not the effect on an animal is likely to be the same in humans. It is well known, for example, that certain strains of experimental rodents are particularly susceptible to cancer, many tumours appearing spontaneously in them. Tests with such colonies would almost inevitably give positive results, the value of which would be seriously open to question. However, scientific judgement and risk/benefit are legislated out by this clause; doubtful results not repeatable by anyone else are legislated in. How the Delaney Clause caused the furore over the cyclamates and how the American consumers rebelled against it over saccharin have been noted in Chapter 4.

One of the most frustrating aspects of assessing tests for carcinogenicity is the level of daily dosage. Test doses are always many times higher and far more frequent than those suffered in human exposure. Some are so high as to cause serious metabolic disruption at a level which in humans would call for hospital treatment and also to raise to significant doses impurities present in otherwise insignificant concentrations. Such massive feeding has focused attention on dose levels of carcinogens generally and raised the question of whether or not there is a no-effect level for them. One line of argument is that no molecules of a carcinogen should ever be permitted; that any concentration, no matter how small, is bad; that there is no no-effect level for carcinogens and, in any case, that the induction time may be so long as to cause one to play safe. As more and more has become known about the mechanism of carcinogenicity at the cellular level, others have argued that theoretically there ought to be a threshold for any carcinogen below which nothing happens or for which there is an induction period longer than any human lifetime. The difficulties of establishing these are compounded by the levels of spontaneous cancer development, variations in individual susceptibility at the concentrations of test substances envisaged and the known protective effect of cruciferous vegetables such as Brussels sprouts, cabbage and cauliflower, and of antioxidants.

One very serious difficulty confronting those who demand the complete absence of carcinogens is the concept of zero tolerance. Nothing found means nothing found by a particular analytical method and as chemical analysis has advanced by leaps and bounds it is possible to measure substances at concentrations of parts per million million (10^{-12}) and so to push zero tolerance lower and lower. As has been pointed out by Bernard Oser, an American food technologist, 57g of peanut butter which contained only 20μg of aflatoxin (see Chapter 6) per kg would contain 2200 million million molecules of this very unpleasant substance. If this concentration were reduced to 1μg per kg, 57g would still contain 110 million million molecules. Thus, there would still be plenty of molecules of aflatoxin present but who would say that that concentration has any toxicological or even carcinogenic significance?

Certainly, the question of the carcinogenicity of food additives and contaminants cannot be treated as simply as the Delaney Clause suggests. On the one hand, far too many substances can be shown to be carcinogenic if the conditions for testing them are made tough enough as in massive continuous dosing, or unreal enough as by repeated injection at the same site. On the other hand, natural substances of known mild carcinogenic properties are consumed by many people without ill effect.

Those of us who have no expertise in this field can do no more than listen to people whose research work is directly concerned with the relationship between food and cancer. In Adelaide in 1978 Sir Richard Doll, the man responsible for the classic studies on the link between smoking and lung cancer, reviewed the subject of nutrition and cancer. Of food additives he said that none had so far been linked to human cancer, nor could he see that the elimination of food additives would be much help in reducing the current incidence of cancer in man. He saw greater risks in the consumption of alcohol (especially if combined with smoking), general over-eating, mycotoxins (see Chapter 6) and lack of fibre in the diet than he could see in food additives. His perception of the risks associated with food additives was similar to those quoted in Tables 5.1 and 5.3. This is reassuring, especially in relation to food additives which have long been in use, but tests for carcinogenicity must always be in the forefront of the testing of substances for use in food.

Mutagenicity: If a living cell suffers a permanent chromosomal change it is said to *mutate*. Other cells which may be derived from it will have different characteristics from it, i.e. they are *mutations* of the original and are called *mutants*.

In nature, mutation is relatively rare but some agents such as X-rays, other radiations and certain chemicals may accelerate mutation. Such agents are called *mutagens* and the capacity for causing such changes is known as *mutagenicity*.

It is not difficult to produce mutants of single-cell organisms such as bacteria and yeasts and some have been deliberately modified to be useful to man as, for example, in improving the yields of selected fermentation processes. Such mutants are desirable economically. What is not desirable is the appearance of mutant cells in man himself. In general, mutations are deleterious to the host and, as a general precaution, there is a strong incentive to exclude from the food supply as far as possible any substance which may be mutagenic.

As has already been pointed out, specific mutagenicity tests in animals are a normal part of the testing programme for substances proposed for use as food additives, but these, and tests for carcinogenicity, take a long time and mutagenicity tests with cells on the laboratory bench have been suggested as short-term tests for carcinogenicity. The first of these was the Ames test, named after the American professor who developed it. It uses a strain of a bacterium called *Salmonella typhimurium* and has been enthusiastically espoused by individuals in various countries. However, it has been shown to give false positives and false negatives and scientific argument still continues about its effectiveness and, particularly, about its significance as an indicator of possible carcinogenesis in man.

There is no doubt that a test taking only a few days to indicate possible carcinogenicity or cell damage would be of tremendous value to every one with an interest of any kind in food additives and contaminants, but the application to man of results obtained on a bacterium is still a matter of dispute. Other tests using mammalian cells have been proposed and offer promise but at present there seems to be no real substitute for tests with animals. Without doubt the rapid tests will be greatly refined and their reliability improved; already they have value in screening compounds for further testing on animals. It is not unlikely that within a few years a combination of two or more of them will form a generally acceptable short method for testing food additives and the results obtained with them will become mandatory for the complete assessment of compounds proposed for use in food or in association with processing.

Allergies and Hyperactivity: Some foods are unsafe for everyone; and, as a general statement, every food is unsafe for someone. The coeliac suffers from a physiological abnormality by which the gluten from cereals makes him sick. The baby lacking a specific enzyme cannot cope with the essential amino-acid, phenylalanine, and grows up mentally deficient. Many Asians have an insufficiency of lactase which digests the lactose of milk and as a result milk products cause them acute distress. Allergic reactions to specific foods are well known, two of the commonest being triggered by milk, which is especially implicated in reactions in infants, and eggs. In both cases sensitivity to a specific protein is blamed. Non-protein foods, such as oranges, also are known to initiate allergic reactions and those who cannot tolerate these foods quickly learn not to eat them. They are entitled to know when they are present in formulated foods and ingredient labelling should enable them to avoid those which cause them distress.

In recent years it has been claimed that practically every type of chemical, especially those derived from coal tar and other hydrocarbon sources such as petrochemicals, are capable of causing allergic symptoms, some of them very distressing, in a not inconsiderable number of susceptible individuals. A second and more specific claim is that 'coal tar' dyes and certain naturally occurring salicylates are responsible for hyperactivity in a high proportion (even up to 25 per cent) of American children. This last is known as the Feingold hypothesis after the man who proposed it.

Now, the general practitioner or allergist faced with a patient suffering from overt and distressing ear, nose, throat and eye symptoms must do something, and there is a growing mass of presumptive evidence showing that the isolation of individuals from a wide range of substances from town gas and petrol fumes to the plasticizers in plastic furniture has a dramatic beneficial effect. Some of the substances implicated are found in foods. For example, one specialist has claimed that a sensitive individual can differentiate between beet sugar and cane sugar, reacting to one and not the other, and this can only mean that the patient is reacting to a small number of molecules carried over from the original plants themselves or from the process and probably present in concentrations well below any chemically detectable level. Other examples, which seem to incriminate some of the additives and contaminants discussed in Chapters 3 and 4, also are known and it has been confidently claimed that *all* chemicals of coal tar or petrochemical origin, i.e. hydrocarbons and their derivatives, have the potential to trigger allergic reactions and that, so far as food is concerned, they should be eliminated or at least declared on the label.

In other words, some foods are unsafe for a small proportion of individuals because they contain contaminants which act in those especially sensitive people at concentrations well below those which have any *toxicological* significance. That this can happen is beyond doubt, but that this can be attributed to substances of coal tar or other hydrocarbon origin is not proven. There is a mass of clinical evidence which clinicians must use to get relief for the patients before them, but there is no rigorous scientific study from which general conclusions may be drawn. Hence, there is no theory, only a hypothesis which is still to be tested. In any case, the active concentrations claimed are so small that there is no way at present either of eliminating the substances from food or of detecting their presence other than by the exquisitely sensitive reactions of the susceptible individuals whose protection lies in eliminating the guilty food from their diets.

The Feingold hypothesis that hyperactivity may be induced in children by 'coal tar' dyes and natural salicylates is in a different category. Given that the hyperactivity has been genuinely diagnosed and that the patient is not simply a spoilt brat, it is possible by means of double blind cross-over tests to put the matter beyond doubt. A double blind cross-over test is one in which a group of patients and a group of controls, in this case children showing no hyperactivity, are given the test substances, i.e. dyes or salicylates, in foods in such a way that they cannot know when they are present and are then given the same foods from which the test substances are absent. By crossing the test and control batches of the foods neither the subjects nor the controls ever know when they are actually under test. The results may therefore be attributed directly to the test substances.

Such tests, carried out exhaustively in the United States and confirmed elsewhere, have led to the firm conclusion that the Feingold hypothesis cannot be substantiated. This is not to say that there will not be individuals who will be exquisitely sensitive to particular molecules – indeed, the reverse is likely – but firm scientific support for the general claim that coal tar dyes and natural salicylates promote hyperactivity in children is lacking. On the the other hand, it seems clear that some individuals react adversely to the yellow food colour, tartrazine. Research continues.

6 Microbial Contamination

'What! Will these hands ne'er be clean?'
Shakespeare, *Macbeth,* Act V, Scene i

There is little doubt that the most serious food-based threat to public health comes not from food additives or chemical contaminants or from naturally occurring poisons in food, but from the microbial contamination of food (see Table 5.3). Chemical considerations depend on the concentrations of a substance present but microbiological discussions centre on the *kind* of organism in the food and whether or not it will grow. This chapter therefore looks briefly at microbes and at the microbiological contamination of food particularly as it relates to the health of the consumer.

Although men like Anton van Leeuwenhoek, who invented the microscope in the late seventeenth century, were aware of the existence of 'animalcules' visible only microscopically, the science of microbiology was developed from Louis Pasteur's work in the 1850s and 1860s and the terms *microbe* and *microbiology* (from the Greek, *mikros* = small and *bios* = life) date from 1881 and 1885 respectively. While the scientist prefers the more specific term, micro-organism, microbe means much the same and certainly the adjective, microbial, is convenient and euphonious.

Everyone is familiar with microbial contamination. Milk turns sour if left for a few days; bread and cheese grow mould if stored carelessly; meat quickly becomes slimy and smelly if left at room temperature. All these phenomena are examples of the results which follow the growth of microbial contaminants.

While microbes may cause much damage and disease, most are useful and some are valuable. Microbes are responsible for the fertility of the soil and for all 'biodegradability', thus ensuring the recycling of animal and vegetable matter in nature.

What are Microbes?

The short definition of microbes is that they are microscopic organisms. They range from viruses, which cannot be seen with the optical microscope and require the special techniques of electron microscopy to make them visible, to bacteria, which are simple single-celled organisms, to yeasts, which are structured more than bacteria, and finally to moulds which form visible threads, called mycelia, from which the 'fruiting bodies' emerge. Both mycelia and fruiting bodies are many-celled but the latter burst to scatter the mould spores, the true single cells, into the air. These are invisible and, if they lodge where moisture, nutrients and oxygen are available, e.g. on the surface of food, they rapidly begin to multiply and soon a mould colony appears.

Other microbes, especially yeasts and bacteria, also are invisible and present in the air, and, unless specific steps are taken to exclude them, are found everywhere; on the hands, on utensils and cutlery, on kitchen benches and on tea towels. They are especially numerous and varied in the soil and in stagnant ponds and pools of water. They grow vigorously on organic matter and hence food is especially vulnerable. They are widely spread, representatives being found in the Antarctic and in hot springs. A few grow in salt brines which would be fatal to most of them. The great majority of microbes are harmless; the few which are harmful to man and animals are called *pathogens,* i.e. disease producing.

Viruses differ in three important ways from moulds, yeasts and bacteria. First, as already noted, they are not visible by ordinary microscopic methods; secondly, they pass through special filters which retain the others; thirdly, they are unable to multiply outside living biological tissue. They cannot, therefore, grow in cooked or processed food, but there is evidence that food and water may carry viruses, e.g. of hepatitis, which then multiply in the body.

There are thousands, if not millions, of different organisms. As already indicated, they fall broadly into four main divisions, moulds, yeasts, bacteria and viruses, but each is subdivided into many families classified according to their morphology (i.e. their form and general structure) and to the way they behave under various conditions. Thus, micro-organisms may be differentiated by their appearance under the microscope, by the characteristic forms of their colonies and by their reactions when grown in solutions of various composition, i.e. in different *media.*

Some organisms under certain conditions form capsules; others have a gummy coating. Under unfavourable, particularly dry, conditions some form spores within the cell. These are quite different from mould spores and survive when the rest of the cell disintegrates. Spores are particularly resistant to heat and, when conditions become favourable again, they germinate and grow back to the original bacterial form.

Many organisms have been named according to their appearance. The general term, *bacteria,* is from the Greek word, *bacterion,* meaning a little rod, and was adopted before it was realized that the 'bacteria' came in many shapes in addition to rods. Now, the Latin word for little rods, *bacilli,* is applied to one particular group, all of which under the microscope look like little rods. Some bacterial cells are spherical and are called *cocci* (Greek, *kokkos* = berry). Some of these form chains, the *Streptococci* (Greek, *streptos* = twisted), or clusters, the *Staphylococci* (Greek, *staphule* = bunch of grapes) (see Fig. 6.1). Others have been named after their discoverers, thus, the *Salmonellae,* sources of much food-borne sickness, were named after Dr Salmon who isolated the first of them from pigs in 1885. Such names as these are family names, but other terms describe the conditions under which groups of micro-organisms flourish and these terms may cover many different families. Thus, some microbes prefer cold; these are *psychrophilic* (Greek, *psukhros* = cold, *philos* = loving). Others which thrive in quite hot conditions are called thermophilic (Greek, *thermos* = hot) and those which grow quite happily in salt solutions are known as *halophilic* (Greek, *halos* = salt).

One of the major sorting tests used in classifying micro-organisms is to establish whether or not the organism requires air (strictly, oxygen) for growth. That is, is it *aerobic* or *anaerobic?* Moulds are aerobic and will only grow if oxygen is available to their spores and mycelia. Hence, food technologists design packages of, say, cheese to exclude air and thus to prevent mould growth. In the few cases in which moulds are deliberately used to produce food, in the preliminary stages of soy sauce production, for example, aeration is promoted.

Some bacteria are anaerobic, *Clostridium welchii* (syn. *perfringens*), for example, which may grow in protein foods with the production of gas. If a can of meat containing an *anaerobe* is underprocessed the organism may grow and produce gas. Therefore, never eat food from a can which has swollen because of gas formation in the contents; it may be a sign of microbial growth.

Some yeasts may grow both aerobically and anaerobically. One ferments sugar to alcohol and carbon dioxide and, at the same time, grows a little, but only a little. This fermentation is anaerobic, but if one wishes to grow yeast, say, for the supply of bakers' yeast, an aerobic fermentation must be used. Air is pumped through the fermenting liquid

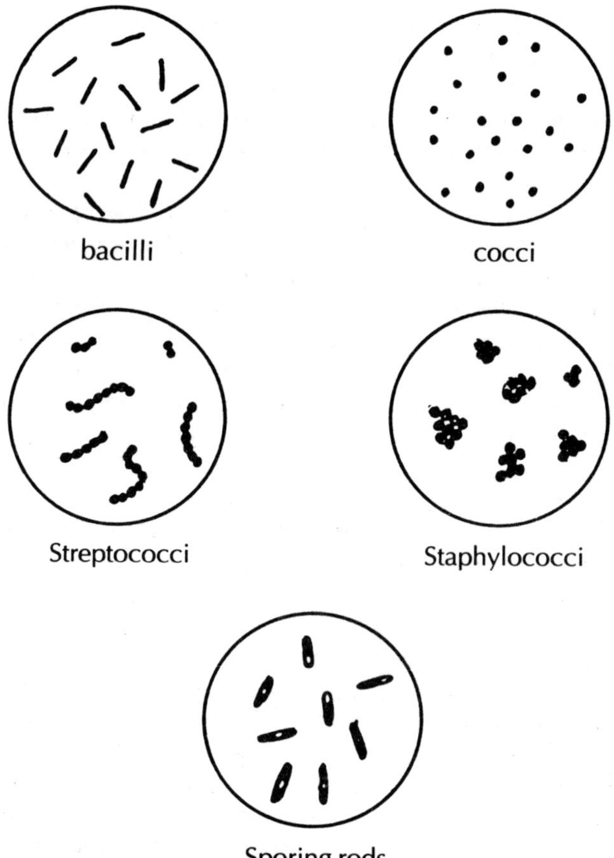

Figure 6.1: Some Micro-organisms

and the main product is yeast cell mass; very small quantities of alcohol and carbon dioxide are produced. The yeasts used for these purposes are specially selected and grown. Many others occur naturally and may infect food, causing unwanted fermentations leading to spoilage. These are called 'wild' yeasts and are contaminants.

Their natural occurrence led man to beer, wine and bread. Other organisms found naturally in milk led to cheese but, in the last 150 years, the microbiological secrets of the methods of making these food products have been unravelled and the organisms required are now grown and used in the pure form. So also with sauerkraut, soy sauce and fermented meat products such as salami. Where organisms called 'starters' are specifically used to initiate a fermentation, other organisms which may creep in are contaminants.

Other groups of microbes are used in fermentations for the production of industrial solvents etc., and even in mining. Such applications of organisms which are beneficial to man are outside the scope of this book, but it should be noted that in every fermentation, as in every food product from which the exclusion of all microbes is desirable, an organism which causes trouble is a contaminant and such contamination should be avoided.

All food raw materials are contaminated and all microbial contamination is to be avoided if possible or at least minimized and controlled. That which is not controlled certainly leads to spoilage and loss of food. It may also make people sick or even kill them through what has come to be called food poisoning. But, as one reads the following pages, it is well to remember that small numbers of most intestinal pathogens, including salmonellae, *Staphylococcus aureus,* the sporing organisms, *E. coli* and, possibly, the vibrios are undoubtedly eaten without ill effects. Food poisoning, as with chemical poisoning, is largely a matter of dose; of pre-formed toxins which act chemically, or of living organisms which must be present in doses large enough temporarily to swamp the body's defences.

Food Poisoning

Early on the morning of 3 February 1975 a Boeing 747 from Tokyo via Anchorage in Alaska touched down at Copenhagen on its way to Paris. Symptoms of nausea, vomiting and abdominal cramps were already evident in some passengers and it soon became clear that the flight could not continue. That morning 142 of the 344 passengers and 1 steward of the crew of 20 were admitted to hospital and the remaining 202 passengers and 19 crew members were kept under observation in Copenhagen hotels. Altogether, 192 of the passengers were affected.

The outbreak followed about one hour after a ham and omelette breakfast had been served and was a classical example of staphylococcal poisoning. Prompt investigation followed and it was quickly found that two cooks at Anchorage had handled the ham. Eighty-six per cent of the passengers who ate the ham handled by one cook who had sores on his hand became sick. There were no cases at all among those who ate ham handled by the other cook. So the origin of the infection was clear and revealed the first mistake. Someone with an infection had been allowed to prepare food.

But that was not all. After infection the ham had been allowed to stand at room temperature for six hours, was then stored at 10°C for 14½ hours overnight and was finally left at room temperature for another 8 hours on the aircraft – an unbelievable sequence of 28½ hours during which the infecting organism had plenty of time to grow. So there was the second mistake: food was allowed to stand about both on the ground *and* in the aircraft at temperatures favouring the growth of micro-organisms, i.e. at temperatures between 5° and 60°C.

Similar episodes have been reported on cruise ships where there is the added hazard of a closed water supply. Incidents involving both faulty catering practices *and* infected water are well documented; and then, some years ago, there was the wedding party which to a man – and woman – ended up in the casualty section of an Australian city hospital in which the bride and groom separately spent their wedding night in lonely discomfort. All were victims of microbial contaminations resulting from faulty food handling by caterers.

Unfortunately, some housewives and home preservers have killed themselves and/or members of their families by faulty heat processing of home-preserved foods. Every now and then someone in the food industry makes a mistake, or a control system breaks down, and some processed food is implicated in an outbreak of food poisoning. This is rare, especially when related to the volumes of processed food produced. It is particularly rare when the food company is supported, as it should be, by a laboratory providing full microbiological surveillance. That many smaller food companies continue to manufacture without such precaution is a serious matter.

Food poisoning, as usually understood, is characterized by the well recognized symptoms of gastro-enteritis, stomach and abdominal pain, vomiting and diarrhoea, and food poisoning organisms fall largely into two groups (see Table 6.1). The first group includes those which poison

Table 6.1: Mode of Action of Microbial Contaminants

Infection	*Production of toxins in food*
Salmonellae	Staphylococci
Vibrio haemolyticus	*Clostridium botulinum*
Escherichia coli	*Bacillus cereus*
Shigellae	Streptococci
	Mycotoxin-producing moulds
Infection to produce toxin in intestine	
Clostridium welchii (syn. *perfringens*)	

by infection and grow in the body to produce symptoms of gastrointestinal distress. It is probably necessary for prior growth in the food to occur so as to build up an inoculum big enough to continue to grow in the body. Examples of this group are the salmonellae.

The second group is made up of those organisms which poison by a toxin which is produced in the food by the prior growth of the microbe, i.e. the poison is already in the food when it is eaten. Examples of this group include the staphylococci, responsible for the victims on the Japanese aircraft. Also included are the mycotoxin-producing moulds but the substances produced by them are so different in action as to warrant separate treatment.

The salmonellae of which there are hundreds of types are gut organisms and are naturally associated with animals and poultry, particularly where they are reared intensively. Salmonellae are always present in abattoirs and slaughter houses and it is therefore difficult to get a frozen chicken, say, which is *not* contaminated with salmonellae. Often unheated animal or poultry feed may be contaminated by these organisms and this ensures the recycling of them. The control of salmonella infection of food raw materials begins with the provision of salmonella-free feed and includes the careful disposal of sewage and wastes. Fortunately, salmonellae are easily killed by heat, even four minutes at 60°C. They are thus absent from heat-processed foods, but the danger lies in non-heat processed foods or in post-processing contamination because they may be transferred easily from raw to cooked products. Thus, the problems are not usually in the factories but in retail outlets, the catering trade and the home.

Vibrio haemolyticus is the most common cause of food poisoning in Japan where it is associated with fish and seafood which is eaten raw. This follows from the Japanese reliance on the sea, and, in many cases, the warmer seas, for food. Food poisoning from this source is worse in the summer months when warmer temperatures favour microbial growth. *Vibrio haemolyticus* has been reported from other parts of the world but its effect is minor compared with that in Japan. It is a cousin of the cholera organism, *Vibrio cholerae,* but it does not seem to be transmitted by water and is easily killed by heat.

Escherichia coli, referred to simply as *E. coli,* is such a common component of the gut flora of man and animals that its detection is used as an indicator of undesirable contamination. It is a common contaminant of many foods and kitchen utensils and certain seratypes are pathogenic. These are known as enteropathic *E. coli* and may cause diarrhoea in babies. Large doses may have the same effect in adults.

Shigellae are similar in many ways to the salmonellae. They are not indigenous to foods but are associated with human carriers and poor personal hygiene. They are most likely to be associated with food handled between cooking and eating, fish, shellfish, salads and the like, but may be transmitted via inanimate objects such as pencils. They affect children more readily than adults.

Man is the major source of staphylococcal infection of foods; acne, boils and nasal discharge especially are potentially dangerous, and people suffering in any of those ways or having any other infection, such as those on the hands of the Anchorage cook, should not be employed in the food industry while the infections last, or they are very likely to contaminate foods pre-cooked and intended to be eaten cold. Since the cause of trouble is a toxin which resists heating, food which has been contaminated may still be dangerous after cooking has killed the causal organism.

Clostridium botulinum is the most feared of all the microbial contaminants of food. It is common in soil and one type is found in the sea and is associated with fish. It is anaerobic and grows well in protein foods to produce probably the most poisonous substance known, a nerve poison. Fortunately, the organism itself will not grow under the acid conditions normally found in fruits or in salted foods. It is fortunate also that this organism is quite easily destroyed by heat. All factory heat-processing conditions are set to kill *Clostridium botulinum* and most cases of botulism are associated with the home preserving of meat and vegetables.

Bacillus cereus is aerobic and it forms spores which may survive cooking and then grow to produce a toxin. This microbe is usually associated with cereals, especially rice and cornflour, though it may also be found in some meat products such as pies. It is a common organism and is known to cause food poisoning which usually occurs in the summer months and is often associated with smörgåsbords where rice and potato may be the vehicles. However, even large doses may cause no harm so it is something of a problem organism.

Certain *Streptococci,* also growing in large numbers, may cause similar problems.

Clostridium welchii is another gut organism and hence is associated with sewerage. It may enter the kitchen via water, vegetables and meat and cause trouble in slowly cooled and/or reheated meat dishes. It must survive the acid conditions of the stomach, probably via the sporing stage, and grow in the intestine to form its toxin. Large doses are required and it is sensible, as a general rule not specifically related to *Clostridium welchii,* to eat food hot or to cool it over 1-1½ hours and refrigerate forthwith.

These, then, are the microbial contaminants responsible for most cases of food poisoning but the emphasis so far has been on contamination coming from the raw material or from the food handler. There are two other routes, post-processing contamination and cross-infection which involve utensils and other equipment used in food handling. The following story illustrates both.

In 1964 there was an outbreak of typhoid in Aberdeen in Scotland and after much scientific work it was established that it came about in this way. A factory in Argentina on the River Plate was canning tongues downstream from a village which emptied all its wastes and sewage into the river. Normally, the factory water supply was chlorinated but one day the chlorinator failed and unchlorinated water was used to cool the cans quickly after processing. One can had a micro-leak in the lid seal and through that tiny leak a minute droplet of water carrying the typhoid infection found its way. This organism grows well in canned meat without causing it to blow or obviously spoiling the contents of the can and when this particular can was opened in an Aberdeen shop and sliced up for customers the typhoid germs had multiplied considerably. They contaminated the slicing blade, dishes used to display other meats, price tags and so on, and the other kinds of meat sliced by that blade and displayed in that cabinet were contaminated with typhoid organisms which continued to grow until the meat was eaten. Now, typhoid organisms are salmonellae which do not produce toxins in food but for which food is a vehicle for the transmission of an infection. So the damage was done, people were infected and many developed typhoid fever.

This episode did three things. First, it suddenly demonstrated the unexpected possibility of micro-leaks in cooling cans, and canneries the world over were advised to use only chlorinated water for can cooling. Secondly, it provided a classic case of post-processing contamination, i.e. the contamination of a product *after* the processing designed to produce a biologically stable product had been completed. Such contamination, in the factory, at the hands of the retailer or caterer or in the home, undoes all the work of the processor. Thirdly, it re-emphasized the dangers of cross-infection, i.e. the transfer of pathological organisms from an infected product to otherwise safe foods, and this has serious implications also for the retailer, caterer and housewife.

Microbes are both ubiquitous and insidious. Unchecked, they are a potential danger both to the food supply and also to public and private health. So, what is to be done? Before looking at the control of micro organisms, we shall consider the comparatively recently recognized special case of mycotoxins, the dangerous metabolites of certain moulds.

Mycotoxins

In August 1722 Peter the Great of Russia assembled a great cavalry army at Astrakhan on the delta of the Volga River. He was preparing an offensive to drive the Turks out of Russia for good. As the squadrons gathered the local serfs began to bring in wagon after wagon full of rye; hay and grain for the horses, flour for the men. Almost at once men and horses went down in agony, smitten by paralysis and a fiery itch, and hundreds upon hundreds of both died. The army was destroyed without even seeing the Turks and that autumn many thousands of people died in the region. The enemy which destroyed the Russian horde and laid waste the countryside was a mould, *Claviceps purpurea,* cause of the Holy Fire also known as St. Anthony's Fire.

The mould, called ergot of rye, produces some twenty poisons as it grows and the disease, ergotism, which results from eating infected rye, is in two forms. One affects the central nervous system, causing paralysis, and the other impedes circulation to the extremities, leading to gangrene. The Holy Fire was a scourge in Europe at least from the ninth century and its cause was discovered by an observant French country doctor at the end of the seventeenth century. Ergotism and its prevention have thus been well understood for a long time – which made the outbreak of the dread disease in France in 1951 a criminal act. Three men for the sake of gain allowed poisonous rye to get into flour shipped to a distant market. Four deaths and much sickness and distress resulted.

The poisons produced by the ergot are mycotoxins, that is, toxic substances produced by moulds, and ergotism is the oldest example of what has become a modern problem. Mouldy foods have been associated in general terms with other kinds of sickness on many occasions in a number of countries over the past 150 years. Specifically, a disease described as alimentary toxic aleukia was associated in Russia with consumption of millet and other cereals which had been snow-covered all the winter and had been invaded by a species of the mould *Fusarium.* Sickness from eating yellow rice discoloured by *Penicillium* species led eventually to the characterization in Japan of a series of mycotoxins, but the sudden surge in awareness of and interest in this class of food contaminant resulted from an economic disaster which hit poultry farmers in Britain in 1960.

A hundred thousand turkey poults succumbed to an unknown liver disease. Some excellent scientific detective work traced the cause of the trouble to some imported peanut meal infected with the mould, *Aspergillus flavus,* and subsequently isolated a specific toxic substance which was named aflatoxin after the mould. Aflatoxin was quickly found to be a powerful carcinogen, specifically the most potent cause of liver cancer known, and immediately the search was on to find out if aflatoxin could be related to liver cancer in man and to determine what other moulds could be adding poisonous substances to human food.

The first question was: is the aflatoxin contamination of food likely in those areas where primary liver cancer in man is common? The answer came quite quickly. Yes, it is very likely. In parts of Uganda and Swaziland where the incidence of liver cancer is high, aflatoxin contamination of food also was found to be high. In 1966-67 4 per cent of nearly five hundred samples of local food in Uganda were found to contain more than 100μg of aflatoxin per kg. This is a very high figure. In Thailand a little later, figures of the same order were found in normal foods and concentrations of nearly 3000 in corn and over 12000 in peanuts were detected. Similar information was obtained from studies in the Philippines. Various foods were implicated and it was suspected that left-overs allowed to stand in the warm humid climate also contributed to the problem.

There was thus presumptive evidence that aflatoxin was associated with a high incidence of liver cancer in man. But there is a lag in the appearance of a tumour following the ingestion of a carcinogen and acute toxicity also was suspected. This was later reported from a number of sources, perhaps the most dramatic being that in western India in which maize containing from 625 to 1560μg of aflatoxin per kg was ingested over a period of some weeks. During this time victims consumed from 2000 to 6000μg of aflatoxin each day and mortality was 20 per cent. Clearly aflatoxin is a most unpleasant substance, but chemists seeking to characterize the molecule soon found that there was not one aflatoxin but several. They were all closely related and all unpleasant, but it was important to find out how they behave in animals which provide human food and if it was possible to get rid of them from food intended for human consumption.

It was soon found that aflatoxin in feed eaten by dairy cattle appeared in the milk. This was one of the chemical variants of the molecule and was given the term aflatoxin M (for milk). It soon transpired that there were two of them and M_1 and M_2 were included in the catalogue of aflatoxins with B_1, B_2, G_1 and G_2 (B = blue and G = green fluorescence in ultra-violet light). These various forms may be differentiated by sophisticated analytical techniques but the total aflatoxin present is what counts. It was also found that they can be removed from foods virtually only by rendering the food inedible. Prevention is not only better than cure; there is no alternative.

In countries where cattle are at pasture the year round aflatoxin in milk is not a problem. An International Dairy Federation survey in 1977 emphasized that aflatoxin in milk could be controlled by limiting aflatoxin in animal feeds, and in 1984 an English study over the two years 1981-83 showed U.K. milks to be largely free from aflatoxin M_1 and the values obtained to be significantly lower than those in milk from other European countries. This satisfactory result was claimed to be due to the effectiveness of U.K. control of animal feeds.

It is well known that warm humid climates favour the growth of micro-organisms of all kinds and, perhaps, moulds especially. It is therefore not surprising to find aflatoxin associated with a large number of food products: cereals, milk, raisins, legumes, sweet potato and cassava – even cocoa and copra – and regulatory authorities everywhere began to introduce very low upper limits for aflatoxin in food, not on the basis of known specific action concentrations but because it was generally agreed that the levels in food should be kept as low as possible. This admirable aim led to difficulties as the following example shows.

In the mid-1960s there was a new awareness that aflatoxin could be a problem in Australian peanuts as in others grown elsewhere. Methods of detection were still being developed but at least one company engaged in the manufacture of peanut butter at once began to look for aflatoxin in the raw material supplied to it. Small concentrations were found but they were so small that it was evident that the matter was not serious. Such concern as there was quickly subsided as the years went by and routine monitoring of the local peanuts showed that aflatoxin was present, if at all, only in barely detectable concentrations, so much so that when the Food Standards Committee in Canberra sought to reduce the permissible limit for aflatoxin in all foods, including peanuts, to not more than 5 μg per kg, industry believed that it could live with it.

In 1977, however, peanuts being received from Queensland by one manufacturer suddenly showed concentrations of aflatoxin of 30, 65, even 100 μg per kg. Toxicologists advising the Food Standards Committee agreed to a limit of 15 μg per kg of aflatoxin in peanuts only and the industry held urgent talks with the Queensland Peanut Marketing Board which took immediate and decisive action. Much of the local crop was condemned and steps were taken to import peanuts from America so that local industry could remain in production. Australian concern can, however, be imagined when it was found that a United States Department of Agriculture 'Nil' certificate for aflatoxin meant 'Less than 25 μg per kg'.

In the meantime, the cause of the 1977 outbreak of aflatoxin contamination of the Queensland peanut crop was shown to be a period of very dry weather followed by heavy rain which caused the nuts to grow so quickly as to split the shells and thus to allow the guilty mould to penetrate and infect the kernel. Fortunately, damaged nuts are easily detected in ultraviolet light and awareness of the permanency of the problem coupled with efficient modern sorting machines has ensured that Australian manufacturers continue to receive supplies of peanuts containing not more than 15 μg per kg. Now that the mould is so well established in the peanut growing area it is most unlikely that the limit of 5 originally desired on *a priori* grounds will ever be attainable; however, government market basket surveys have shown that the limit of 15 is being met.

Whether or not such variations in aflatoxin concentration are significant has already been discussed (p. 103).

While aflatoxin has attracted most attention, other mycotoxins also are known; sterigmatocystin and ochratoxin A from other species of *Aspergillus* and several from various *Penicillium* species are examples. There is no doubt that a great deal remains to be learned about moulds and the toxins they produce. In the meantime there is every reason to be more careful with mouldy foods than we have been in the past. Spots of mould should be cut out with a generous amount of apparently unaffected surrounding food and thrown away. Badly moulded food should be discarded altogether.

Microbes Controlled

Without control the primary objective of food technology, the conquest of time and space by preservation for storage and transport, is lost. It is lost, as we have already seen, by the activities of microbes in rendering food either inedible, or dangerous, or both.

Control begins with the raw material. Heavy contamination may require excessive processing but, in any case, is likely to ensure serious quality defects in the finished product. It is a fallacy to imagine that processing will overcome raw material deficiencies. The resulting product may be perfectly safe, but it will undoubtedly be of poorer quality. So close attention to cleanliness and temperature in the production and storage of raw materials is the first step in control.

Raw materials include water: water as an ingredient, water for washing and for cooling. Water is a common vehicle for microbial contamination which is why it is chlorinated, the very low risk associated therewith (see p. 60) being as nothing compared with the benefit of having it pathogen-free.

Table 4.2 (p. 76) shows the processing methods used to control organisms. Fresh food may be frozen or refrigerated but almost all microbes will survive. Cold simply reduces growth and multiplication to a minimum. It must never be forgotten, therefore, that as soon as frozen food thaws and, with refrigerated food, begins to warm up, the micro-organisms which are always present will begin to multiply and, given time for growth, spoilage will follow.

Drying, salt and sugar prevent the growth of microbes by denying them water but, as soon as food is reconstituted or diluted, they begin to grow and the food must be cooked or heated to kill them or consumed at once. Exclusion of air, as we have already seen, will prevent the growth of moulds and some other organisms, and preservatives, as we have seen in Chapter 4, selectively kill or inhibit spoilage organisms and some pathogens.

Heat processing is designed to kill microbes and is the most positive way of controlling them. Pathogens, but not their spores, are killed at pasteurizing temperatures, say, 65°C or above for varying times; articles may be sterilized by pressure cooking and canning kills almost everything because the conditions of time and temperature are based on those required to kill spores of *Clostridium botulinum*. These conditions will kill other organisms, too, and modern factories work with recorder/controllers on the heating equipment so that the temperatures required are actually attained and the temperature history of every batch is known. This is one reason why poisoning from factory heat-processed food is so rare – if the process is working properly, the organisms are killed.

Some preserved foods, salami, for example, are not heat-processed and rely on lowered effective moisture content to prevent the growth of contaminating microbes. Most of the time this kind of product is satisfactory, especially if supervised by a qualified microbiologist, but occasionally trouble appears.

It goes without saying that regular and thorough sanitation of all plant, utensils and equipment is essential and that it must be backed by regular and thorough inspection, and supervision of the personal hygiene of plant operatives. The quality of both water and air supplies also must be kept under review. Regular and on-going storage of samples of final product followed up by microbiological testing after specified times at specified temperatures is the most positive way of monitoring the safety of factory procedures. Unfortunately, such testing is slow. Time must be allowed to enable the organisms, which one hopes will not be there, to grow. Consequently, a system called Hazard Analysis Critical Control Points (HACCP) has been developed in the United States. This system aims to establish both the points of greatest microbiological hazard in a process and also the critical parameters which must be satisfied at those points to ensure a finished product which is microbiologically stable and therefore acceptable. The control at those critical points may then be exercised by ensuring the achievement of a standard predetermined microbiologically. Tests, such as the measurement of temperature, the time at a given temperature, pH, water activity or moisture content may then be applied in the knowledge that if the standards laid down are achieved, then the final product will be sound. This is not, in fact, a new system at all but a new name for the commonsense application of microbiological principles.

From the factory viewpoint protection against post-processing contamination means one of two things. First, where the product is packed hot, the integrity of the package. Secondly, where the product is packed at lower than processing temperatures, exclusion of microbes as well as the integrity of the package. Special precautions must be taken where food is packed cold and after processing; the Aberdeen typhoid outbreak is sufficient example of the importance of package integrity.

When a commercial product is implicated in a case of food poisoning, the company, if it is properly organized, swings into a product recall procedure, the essence of which is publicity. Consequently, everyone hears about it. Similarly, if a large number of people are affected after some private or public function, the size of the outbreak guarantees publicity. What is rarely publicized or even recognized is the frequency of mild or even severe food poisoning which affects individuals or families. Food poisoning is not a notifiable disease and if the symptoms clear away within a day or two there is no follow-up. This is because follow-up is messy, involving the microbiological examination of the stools and vomit of the victim and of all suspected foods; it is also time-consuming and expensive. Where an attack is mild or affects few people, lack of any attempt to explain it is at least understandable, but because of this lack the full impact and potential danger of careless food handling are not appreciated. Food poisoning should be a notifiable disease everywhere as it is in some countries already.

Undoubtedly, the major problems with the microbial contamination of food arise from carelessness in the handling and/or holding of it at temperatures favourable to microbial growth. The incidents already referred to make this clear. It is also clear that retailers, caterers and the housewife have heavy responsibilities in assuring the wholesomeness of food.

It cannot be emphasized too often that the potential of food to cause sickness depends on the kinds of organisms present and whether or not they will grow. No food-poisoning microbes except moulds, which may produce mycotoxins but which quickly become visible, can grow in acid foods and, therefore, most fruits and several fermented products may be considered to be safe. Because pathogens are most frequently associated with animals, animal products are more likely to initiate outbreaks of food poisoning and meat and egg dishes, therefore, require special care. However, vegetables may be contaminated in the field and all food is liable to contamination via an infected water supply or by utensils, cloths and benches in the kitchen. If it is accepted, then, that all food, even in the cleanest looking kitchen, is contaminated by something or other, what must be done to prevent growth which would lead to trouble? Fortunately, there are clear guidelines which translate into a simple system of Dos and Don'ts.

Growth is inhibited above 60°, refrigeration (5° or less) will slow down growth to a minimum and freezing virtually stops it. Between those temperatures growth is likely and in the middle of the range it is certain, so some authorities require prepared food to be held by caterers and retailers at temperatures below 5° or over 60°C.These regulations are designed to ensure that organisms present do not grow. The same precautions should apply in the home where the housewife can steer a safe course amid the shoals of microbial contamination by following these imperatives:

DO

Keep food only below 5° and above 60°C.

Eat food hot or refrigerate as soon as possible.

Wash hands thoroughly before preparing food and after handling raw meat, fish and poultry. Keep flies out of the kitchen and certainly off food.

Clean all utensils thoroughly after preparing raw meat, fish or poultry.

Keep pet foods and all utensils, knives, spoons, etc. used for raw pet food completely separate from domestic food, dishes, tea towels etc.

Reheat 'take-away' or previously cooked food right through to at least 75°C.

Ensure that frozen meat or poultry is fully thawed before cooking otherwise heat will be absorbed in thawing the residual ice and the interior of the meat will not get hot enough to kill the contamination inevitably present.

Keep kitchen and all utensils free from all signs of visible dirt and grease.

After use, rinse all cloths and mops in clean hot water and spread them out to dry: cloths left damp and bunched are ideal breeding grounds for microbes. For the same reason, spread tea towels out to dry.

When discarding food, wrap it up before putting it in the garbage can – otherwise the flies will find it and spread contamination.

DON'T

Handle food when suffering from an infection.

Handle cooked and uncooked meats together.

Store cooked and uncooked meats in contact with each other.

Eat food from a can which is swollen.

Forget that contaminated food may look and taste normal.

These rules are derived from basic principles which are part of the armoury of every food technologist. They should be memorized by everyone employed in the preparation and selling of ready-to-eat foods, hot or cold, and in the catering trade.

7 Regulation and Control

> The great want in this colony . . . is an authorised system of supervision over the articles of food and drink retailed for public consumption.
>
> *Argus,* Melbourne, 5 October 1881

As we have already seen in Chapter 1, concern for what was being added to food surfaced in Britain in 1820 but did not become a public issue until the middle of the century. Emphasis was on adulteration as a form of deception. In 1858 some two hundred people were poisoned in the city of Bradford in Yorkshire and seventeen died. The cause was a batch of lozenges accidentally adulterated with arsenic instead of the plaster of Paris normally (!) used to 'extend' the mixture. This episode ensured the passing of the first British law specifically to prevent the adulteration of food and drink.

Analytical chemistry was then in its infancy and the detection of adulteration was not easy. It is difficult for us today to understand how hard it was in those days, even when the press was vigorous in denouncing adulteration, for governments to regulate what was added to food, to control food in the market place and to gain convictions in the courts for offences against what laws there were. The science of the day was not good enough to provide the evidence necessary and the organization for regulation and effective inspection was non-existent or inefficient.

Things began to change in the 1870s. In England analytical methods for the examination of food were developed and published but there was no definition of adulteration and the inspection service which was set up was not compulsory. The authorities relied on prohibiting the sale of food 'to the prejudice of the purchaser' or of any food 'not of the nature, substance or quality demanded by the purchaser'. This philosophy transferred easily to British colonies but it was not specifically additive-oriented; water in milk and alum in bread were special targets.

Regulation

Reference has already been made (pp 20-1) to related problems in Australia involving colours and preservatives. They led in 1905 in Victoria to the first *Pure Food Act* with its associated Regulations and provisions for the regular sampling and analysis of food offered for sale. Because most health matters had been left to the individual states of the new Commonwealth at federation on 1 January 1901, Victoria's Act had local significance only, but the other states quickly followed and by 1912 all had similar but, unfortunately, not identical, Acts and Regulations. New Zealand had done away with provincial governments in 1876; consequently, its *Sale of Food and Drugs Act* of 1877 applied evenly throughout the country. The philosophy in both countries was 'prohibitive', excluding from food those substances not specifically permitted to be added.

In 1874 Canada's *Internal Revenue Act* prohibited the sale of food containing any injurious or poisonous substance. *The Adulteration Act* of 1884 included four important principles related to food additives:

- (a) prohibition of the addition of anything to food which may injure health
- (b) provision for tolerances
- (c) labelling requirements
- (d) exemption of certain ingredients from the labelling requirements

Points (b) and (c) were especially important and (b) farsighted also. An amendment in 1890 to *The Adulteration Act* established the principle of standards of quality and composition of foods.

As in other countries so also in Canada there was in the 1890s public concern over preservatives and colours. This resulted in a detailed study by Dr A. McGill of preservatives and colours, especially the former. It was published in the *British Medical Journal* in 1906 and his report to the Canadian Government (Bulletin No. 126 (1906) of the Department of Internal Revenue) contains six recommendations which are worth quoting in full because, although directed specifically at preservatives, they represent the first enunciation of the principles governing the use of food additives. They are:

- (a) the harmlessness of a preservative should be demonstrated unequivocally.

(b) the general principle that every new thing must be required to show cause for its existence should be applied to the use of preservatives in foods.
(c) the very potency of preservatives, by virtue of which they can be employed in small amounts, is a reason for limiting the amounts to be used.
(d) careful study of any new substance offered for trial as a preservative should be made since it is not necessary to risk a trial with a new preservative in food which shall involve possible injury to health.
(e) everything in the nature of a food preservative should be illegal for use unless specifically permitted, with due consideration to be given to such items as sugar, salt, vinegar, wood smoke and alcohol as permissible without limitation.
(f) it is right and a necessity that the presence of a preservative and its name always be announced on the label.

These principles call for toxicological safety, technological need, labelling and, in the absence of permission to include, exclusion. They are, in effect, those which apply to food additives generally today.

McGill's recommendations led to a standard for milk and to the prohibition of the addition of certain named substances to vinegar and to alcoholic beverages. The Canadian *Food and Drugs Act* of 1920 provided for the making of regulations by the Governor-in-Council.

In the earliest days British attitudes to food laws had understandably crossed the Atlantic to the American colonies which, in due course, began to enact their own local laws. As elsewhere, the concern was for the protection of the consumer from adulteration of the food supply, and, after the Revolution, the individual American states continued to take what steps they deemed to be desirable. Some people, however, saw value in central control and a trade journal held a competition for a suitable Act. The winner, announced in the December 1880 number of the journal, was G. W. Wigner, a leading British public analyst and, at the time, the honorary secretary of the Society of Public Analysts. He proposed, *inter alia,* a system of boards of health appointing their own public analysts, a system he was used to, but a bill based on Wigner's ideas failed to get through the United States Congress.

Many more bills were introduced in the next twenty-five years but State rights and lack of public conviction delayed until 1906 the passing of the *Foods and Drugs Act*. There is little doubt that even then it was the activity of H. W. Wiley and his 'poison squad' which got it through – and Wiley's volunteers ingested large doses of additives!

This law stopped the use of many harmful substances but it was the period of the establishment of the GRAS (generally recognized as safe) list. The 1906 Act was superseded in 1938 by the *Food Drugs and Cosmetics Act* but there was still no requirement that a pesticide or food additive should be shown to be safe before it was used. The regulatory authorities had to prove to the court's satisfaction that a substance was harmful as used or in food as sold. The onus of proof was on the government and it was difficult to accomplish.

Preoccupation with English speaking countries may be understandable but it may also be less than fair to others. For example, in France the statute of the pastrycooks of Bourges of 12 May 1574 forbad the use of saffron lest it be thought that eggs had been used, and in 1641 the city fathers of Amsterdam prohibited the addition of annatto or other colour to butter to prevent the pale winter product from being sold as the more popular deeper yellow spring butter, which, incidentally, would have been much higher in vitamin A activity. In both these examples the use of a specific additive to deceive was forbidden.

In 1834 the Danish government noted the use of harmful colours by confectioners and pastrycooks and on 21 December 1836 the Copenhagen Commissioner of Police issued a permitted list of colours for food and toys. It was probably the first such list anywhere and in April 1843 it was extended to the whole of Denmark. It contained 39 colouring materials, mainly natural extracts, and it is worth noting that it was issued well before the first synthetic dye appeared. By contrast the German *Colour Act* of 1887 prohibited the use of harmful colours in foods. In both these cases concern was for the health of the consumer.

In Australia the Victorian *Pure Food Act* of 1905 provided for the appointment of a Food Standards Committee with power to make regulations and the first regulations under the Act related to food additives; preservatives and colours. It was concern over these, especially the former, which had led inexorably to the Act itself as, of course, had been the case in America and Canada. In 1925 Britain, following the report of a government committee, took specific action on preservatives but a more lenient view of colours, simply prohibiting some. Other countries also acted in the earlier part of this century but public concern seems to have been overshadowed by two world wars and the troubled years between.

Things changed in the 1950s and they changed for the reasons already stated in Chapter 1; a new awareness of the environmental implications of cancer, dramatic advances in the methodology of analytical chemistry which made contaminants detectable and measurable, and new needs and opportunities for the international regulation of foods.

In 1947 the United Kingdom established a Food Standards Committee to advise Ministers on the regulation of food. Membership included government officers, academics, the food industry and public analysts. This move followed a pattern which had been established in Victoria in

1905 and copied by the other Australian states in the succeeding six years. It is an excellent system based on scientific understanding, technological practicality and consensus and in which any exaggerated ideas industry may have are moderated by government, academic and, now, in Australia at least, consumer representation. It works!

The United Kingdom Food Standards Committee immediately set up specific sub-committees, for example, a Metallic Contaminants Sub-committee in 1948 and a Preservatives Sub-committee in 1951, but up to the mid-1950s there was in Britain only a short positive list of preservatives which could be added to foods and a list of colours which were *not* permitted to be added; which is not to say that industry was irresponsible. In fact, the major suppliers of food colours, for example, also maintained extensive programmes of testing on animals and the range of colours offered to the food industry was therefore very limited. The *Food and Drugs Act* of 1955 provided for controls over additions to foods and by 1964 the various sub-committees had evolved into the Food Additives and Contaminants Committee of the Ministry of Agriculture, Fisheries and Food. The reports of this committee, which is an expert committee of advice, are open for public comment, have made valuable contributions to the overall body of information on the subject, and have influenced the making of regulations on food additives and contaminants in other parts of the world.

In Australia the fragmented system of food regulation is a legacy of colonial suspicions of central authority at federation in 1901, but in 1953 the Australian Commonwealth Government's Department of Health set up a Food Standards Committee and a Food Additives Committee to guide State authorities by providing advice, which rarely goes unheeded, on what may be added to or permitted in foods. The Food Additives Committee, an expert committee, concerned itself immediately with food colours and continues to this day, under another name and with a long history of achievement behind it.

Canada and New Zealand, with organizational variations appropriate to those countries, work in similar ways. The important thing is that in all four of the countries mentioned advice on additives, whether channelled through Food Standards Committees or proferred directly to Ministers, emerges from groups of experts responsible to none but themselves and arriving at their conclusions on the scientific evidence free from sectional or political pressures.

As trade relations between Australia and New Zealand have become closer, so the incentive to collaborate in food regulations, already very similar, has increased. It is a state of affairs which, on a much larger and more complex scale, has been faced in Europe. There was an early European *Codex Alimentarius* and, since 1962, the EEC has been issuing directives on additives which have culminated in the allocation of specific numbers, the E numbers, for each additive permitted and the requirement that the name of the additive or the E number must be shown on the label of every food in which it is included. It is a good solution to an awkward problem and Australia and New Zealand adopted the same system from 1 January 1987.

The United States has its own problems. The shortcomings of the 1938 Act were tackled after the war and the Pesticide Amendment of 1954, the Food Additives Amendment of 1958 and the Colour Amendment of 1960 followed. The constraints of the Delaney Clause in the Food Additives Amendment have already been discussed in Chapter 5 but the GRAS list also is a source of difficulty. The GRAS (generally recognized as safe) list of additives is a long one. It was hallowed by time and, with the permissive philosophy which allows the wide use of substances on it, is the source of almost all the adverse news media publicity which is directed at the food supply – any food supply regardless of the regulatory philosophy of the country in which the publicity occurs.

Those who 'generally recognized' a substance as safe were experts qualified to do so, not the Food and Drug Administration; accordingly, in recent times there have been reviews of the GRAS list and such reviews, given the length of the list and the peculiar circumstances of the American scene, became very political, involving lawyers, vocal consumer groups and issues far removed from the scientific facts. The only way to handle the regulation of food additives and contaminants is *on the scientific evidence* – which is subject to change as the boundaries of knowledge extend.

The scientific evidence available to regulatory authorities comes from all over the world via governments, United Nations agencies, industry, research institutions, the scientific and technical literature and personal communications from individual scientists and scientific attachés. The sources are so varied as to make it easy for claims and statements to be cross-checked. Much of this cross-checking and overall assessment is done by the Joint FAO/WHO Expert Committee on Food Additives (JECFA) which was set up in 1957 and at once adopted a set of principles by which it has acted since. They are as follows:

- Food additives should not be used to disguise faulty processing or handling techniques, nor to deceive the consumer with regard to the nature or quality of food;
- Special care should be exercised in the use of additives in foods that may form a major part of the diet of some sections of a community, or that may be consumed in especially large quantities at certain seasons;

- The choice of food additives should be related to the prevailing dietary patterns within a community. The availability of essential nutrients and their distribution in the various foods consumed should be taken into account before the true significance of making a further addition of a particular nutrient (e.g. calcium or phosphorus) or of using an additive that may change the pattern of nutrients in a food (e.g. an oxidizing agent) can be assessed;
- The specifications needed for each food additive have been compiled with three main objectives in mind:
  to identify the substance that has been subjected to biological testing;
  to ensure that the substance is of the quality required for safe use in food;
  to reflect and encourage good manufacturing practice.

These are good principles and, taking a world view, draw special attention to the need to watch for over-concentration on certain foods at certain times – for religious reasons, say, or because of seasonal scarcities of other articles of diet. They take particular note also of the possibility that certain additives may skew a diet in the wrong direction. Further they may be said to emphasize those other general principles which apply to all aspects of diet, both positive and negative – moderation in all things and as much variety as possible – principles which sit easily with those advocating them in well-fed societies but which have a hollow ring when enunciated in communities where the primary concern from day to day is simply to get enough to eat.

Overall, however, they recognize the place of food additives and safeguard the consumer. They acknowledge three things: technological necessity, public health responsibilities and protection of the consumer from deception. The consumer may be protected from deception by appropriate labelling regulations so that the two matters on which the regulatory authority must be satisfied are technological need and toxicological safety.

The information generally sought in each submission made for the introduction of a new additive includes a full chemical description of the compound and its properties; its method of manufacture and likely impurities; the foods in which it is proposed to be used, the concentrations proposed and the likely daily intake by both children and adults; why it is to be added and its advantages to consumers; its status in other countries and any standards of purity imposed; its stability in food and drink; and, finally, detailed toxicological evidence from at least two kinds of animals. These requirements are heavily weighted towards establishing safety-in-use but technological need also must be demonstrated.

There are instances on record of additive submissions failing on one count when the other has been satisfied; of the denial, for example, of the extension to another product of an additive already permitted in one product because the technological need for it in the second had not been demonstrated to the satisfaction of the committee; and, of course, of the failure on toxicological grounds of submissions for additives for which technological need had been demonstrated. The mere fact that a substance has been judged to be toxicologically safe does not mean that it will necessarily be permitted in a particular product or in any product at all for that matter. Technological need for the specific use must still be demonstrated. But there is another aspect; technological need may have already been established for, say, an emulsifier in ice-cream or mayonnaise and someone may then want to introduce another emulsifier. Some may say that there are already enough emulsifiers in the list and that another should not be permitted, but the regulation-makers have no brief for supporting those types or brands of emulsifiers already in use. Further, permission to use a new toxicologically acceptable compound will lead to an overall reduction in the use of those already permitted and hence contribute to that variety already noted as being desirable.

Toxicological safety has been discussed in Chapter 5 and, as was pointed out there, the information available is used to set upper limits for food additives and contaminants in food products and raw materials. These are regulations, not safety limits. The latter will be many times higher than the former for we have seen that in doing their sums, toxicologists use a safety factor of a hundred. The quite frequent journalistic ploy of reporting that something or other has been found to exceed the safety limit by two or three times is therefore nonsense. It is possible to exceed the regulatory limit two, three or even ten times, except in isolated cases, without reaching the safety limit. It is always an offence to exceed the regulatory limit, but, as we have seen, public health is rarely at risk from either additives or contaminants. On the other hand, knowledge never stands still and new information is emerging every day to be taken into account in the regulatory process. Consequently, we should not expect regulations to remain static: some will be tightened; some will be relaxed; but in today's atmosphere of caution and watchfulness it is unlikely that anything of significance will be overlooked.

It may be as well to remind ourselves, however, that from some points of view the very tight food regulations demanded by western communities are a luxury largely confined to the developed nations. Thus, FAO and WHO recommended a limit of 30 μg of aflatoxin per kg of food because those making the recommendation knew full well that the danger of malnutrition through insufficiency was greater than the danger of liver cancer. The subjectivity of the basic dilemma on what to some seems a clear-cut scientific/regulatory issue has never been better expressed than by Dr M. O. Moss writing in 1975 in the *International Journal of Environmental Studies:*

> 'I wonder how often an epidemiologist has had to leave his field studies, to write up the results, in the knowledge that such and such a family has consumed, and may continue to consume what the legislative machinery in another part of the world considers to be a dangerous level of a known fungal toxin.'

Such a statement emphasizes that developing countries have their own problems of contamination, some naturally occurring, some industrial. They also have real needs for agricultural chemicals, specifically, pesticides, not only to maximize the production of crops, but also, and perhaps even more importantly, to protect those same food crops between harvest and consumption. Pesticide usage is only one, but an important one, of several methods available for the reduction of the enormous post-harvest losses of food in the Third World, and the limitation of the residues is as desirable and necessary as in the more developed countries. Unsurprisingly, the Third World turns to *Codex Alimentarius* for guidance and regulation.

The Codex Alimentarius Commission was set up in 1962 as a Joint Commission of FAO and WHO. It is representative of the governments of the member nations of those two organizations and the meaning and scope of *Codex Alimentarius* is clear from the guidelines of the Commission which say, *inter alia,*

> 'The *Codex Alimentarius* is to be a collection of internationally adopted food standards presented in a unified form. These food standards aim at protecting consumers' health and ensuring fair practices in the food trade. Their publication is intended to promote the standardization of foodstuffs in various parts of the world, to facilitate harmonization of standards, and in so doing to further the development of the international food trade.'

Since 1962 the *Codex Alimentarius* has followed these guidelines consistently. Many standards have been written and either adopted outright or used by emerging countries in the framing of their own regulations. Not the least of *Codex Alimentarius'* contributions has been the work on additives and contaminants.

The major problem in the regulation and control of aditives and contaminants in the less developed countries is the very high proportion of food produced and, certainly in the ASEAN countries, increasingly being packaged in the villages. As the author was told in 1985 in the Philippines, where some 70% of the people still live in the countryside, 'We have no idea what is being used to colour and flavour food produced in the villages.'

In the cities it is easier. To the author's personal knowledge, there are in Manila, Kuala Lumpur, Singapore and several Indonesian cities large food processing plants of world standard. Some are locally owned and operated and what they do is very visible. Others are branches of international companies working to their own internal standards but conscious of their reputations. They know that what they do *there* may rebound on them *here*, so to speak, and they all have clearly before them the classic case of the large and well known international company which incurred obloquy not because the product was poor or suspect, but because of the social consequences of the promotion of it. In addition, of course, those in such countries who export food products to western markets are well aware of the tight regulations which they must meet in them.

Control

It is one thing to have regulations; it is another to police them. The British method from the middle of last century was to charge local authorities with the task of enforcing the law relating to food and this was duplicated in some overseas territories. Today, many governments prefer a centralized system of food supervision, but, no matter how they are taken, food samples are analyzed.

The preparation of a sample for analysis usually follows a defined procedure and the method of analysis also is often specified, or even defined, in the regulations. Analysis for the presence of chemical substances, say, a heavy metal contaminant such as cadmium or lead, is straightforward compared with microbiological examinations. Atoms of elements or molecules of compounds are so small that even if they are not distributed uniformly throughout a sample in the first place, rapid mechanical mixing will quickly ensure that a sample is rendered chemically homogeneous for the purposes of analysis and the results will be expressed simply as mg of cadmium or lead per kg of food, or, in the case of the major constituents, as percentages of water, fat, starch, sugar or protein.

The microbiologist is concerned with numbers, as is the chemist, but has a much more difficult task. Microbes are much bigger than even the largest molecule and no amount of mixing will make a sample homogeneous with respect to microbes if the organisms are present in small numbers. In any case, they will be scattered through a sample like currants through a plum pudding. If the currants are plentiful each slice will have much the same number, but if they are few they are likely to vary from slice to slice and it may be possible to get a slice without any at all. If this happened, however, it would not be possible to say that the cook had left them out altogether without taking more slices or, perhaps, even eating the whole pudding.

So, also, in sampling foods for microbes, especially those which are likely to be dangerous or which are indicators of potentially dangerous contamination. Accordingly, methods of sampling based on the principles of mathematical statistics have been devised and standards relating to them are being drawn up. Such a standard will say something like this: there shall be no organisms in four out of five samples tested and not more than so many in the fifth. This makes sure that any microbes present are few in number and that any 'clump' is not a large undisturbed focus of infection.

This system is designed to reveal the presence of single organisms, but the argument on which it is based applies equally to aflatoxin in a batch of peanuts (see p. 120). A relatively few nuts infected with the mould and high in aflatoxin may be mixed through an otherwise clean consignment. This led a confectioner putting one nut on each chocolate to ask at the time of the aflatoxin scare some years ago if he were playing Russian roulette with his customers. The answer was, Yes. Special sorting techniques have since greatly reduced this risk but, because the production of aflatoxin derives from a microbial infection and is hence localized, routine sampling of consignments of peanuts follows a statistical procedure similar to those used by microbiologists in the sampling of foods generally.

Both chemical and microbiological analyses are costly. This is especially so of microbiological examinations because of the time taken for organisms to grow, and when food additives and, especially, contaminants which occur in such low concentrations, are concerned, it may be *very* costly. Local authorities are therefore chary of being involved in the finer points of the control of the food supply. Even when central government laboratories undertake this work, the number of samples which can be analyzed in a given time by a staff of a given size is limited.

Some twenty years ago it became obvious that the rapidly increasing number of regulations relating to food additives and contaminants would be worthless without the concomitant policing of them. The attention of the relevant authorities in at least one country was therefore drawn to the need to expand both the laboratories and the staffs and to purchase the new equipment needed. Nothing happened!

Now, as one government analyst told the author of this book, regulations are to keep honest men honest and there is no doubt whatever that this happens. The large food companies, especially, have good technical support; some are first class. All of them go to great lengths to observe the regulations, for the only thing feared more than an adulteration or similar charge is a microbial problem leading to product recall. That is even more public. But there are many small companies which, even today, neglect or skimp on laboratory control and rely entirely on what technical salesmen tell them about food additives. Usually that is correct but sometimes it is not. Nor do all imported foods conform with local regulations. All in all, there is a case for the expansion of government food surveillance laboratories. Indeed, people who are sometimes very vocal about what they think is wrong with the food supply should instead be demanding a more extensive application and policing of the very good regulations which are already in place.

This will, of course, cost money and the whole question of the surveillance of the food supply in general and of food additives and contaminants in particular comes down to this:

How much is the community prepared to pay for its protection?

Glossary

accelerator, particle: a device used to accelerate sub-atomic particles, such as electrons, to high velocities

additive: see food additive

adulterant: in foods, an adulterant is a substance foreign to the food and added illegally for the purpose of gain without the knowledge of the purchaser

aerobe: an organism requiring air (i.e. oxygen) for growth

aflatoxin: a very toxic and carcinogenic substance formed by the mould *Aspergillus flavus* and found in foods when that mould grows on them

aliphatic: open chain organic compounds derived originally from fats

alkaloid: a complex compound of carbon, hydrogen, oxygen and nitrogen found in plants and having important physiological properties. Many, such as coniine from hemlock, are very poisonous, others like cocaine are addictive, and some, when used in small doses, are valuable in medicine

allergen: a substance capable of promoting allergic reactions in individuals

amine: a compound formed from ammonia and a hydrocarbon radical

amino acid: a compound, usually aliphatic, containing both an amino ($-NH_2$) and a carboxylic acid ($-COOH$) group. Proteins are built up from amino acids

amino group: a combination of one nitrogen and two hydrogen atoms derived from ammonia

anaerobe: an organism which grows only in the absence of air. Aerobes which can adapt to grow in the absence of air are called facultative anaerobes

analogue, meat: a product made, usually from vegetable protein, to simulate meat in texture and flavour

anthelmintic: a substance used therapeutically and prophylactically to protect animals from worms antioxidant: a substance which protects fats from oxidation, i.e. rancidity

aromatic: in chemistry, aromatic compounds are those derived originally from benzene, a ring of six carbon atoms; the term now includes most organic compounds having ring structures

biodegradable: a term used to describe materials which are capable of being decomposed naturally, i.e. by air, water and the microbes and other organisms present in the natural environment

buffer: the name given to a compound which has the property of maintaining the pH of a system relatively constant despite the addition of either acid or alkali

carboxylic acid: an organic acid in which the acidic group consists of one carbon, one hydrogen and two oxygen atoms in the form –COOH

carcinogen: a substance which causes or promotes cancer: hence carcinogenic

cariogen: a substance which causes or promotes dental caries: hence cariogenic

catalyst: a substance which promotes a chemical reaction without itself taking part in it

contaminant: a chemical contaminant in a food is a substance which is not normally present in that food in its natural form or which is present in concentrations not normally found or which is not permitted under the food regulations to be present, or, being an additive as defined under the regulations, exceeds the concentration permitted

DDT: 1,1,1-trichloro-2,2-di-(4 chlorophenyl) ethane; the term is frequently used to include this and four closely related compounds known as DDE, TDE (or DDD) and methoxychlor

detoxifying mechanism: a biochemical reaction of series of reactions which operate in the body to overcome the toxic effects of a substance

enzyme: enzymes belong to a special class of proteins produced by living cells and have the property of catalyzing specific biochemical reactions in plants and animals. They are vital to all life processes but are destroyed (inactivated) by heat and by chemicals which precipitate proteins. Most are named by adding -ase to the substrate which they hydrolyze (e.g. proteinase, maltase) but some retain their old names (e.g. emulsin, pepsin, rennin)

epidemiology: the study of disease and its causes on a geographical or community basis

essential oils: a term used to describe the characteristically fragrant oils recoverable by steam distillation from the leaves, flowers and fruits of plants

food additive: a substance deliberately added to food by the manufacturer to facilitate processing or to improve appearance, flavour, texture, keeping quality or nutritional value

glycosides: compounds of a sugar with a non-sugar

histamine: an amine found in tissues and released when they are injured: it causes dilation of the blood vessels and leads to local swelling

hyperactivity: abnormal pathological activity

hypertension: essentially, high blood pressure

hypervitaminosis: poisoning due to excessive intake of vitamins

hypoglycaemia: a deficiency of sugar in the blood

humectant: a substance added to foods and other commodities such as tobacco to keep them moist

ingest: lit. to carry food into the stomach; having also the connotation of absorption

inorganic: compounds of all elements other than carbon

iodophor: a name given to a group of compounds containing iodine and which give it up from solution to act locally as a disinfectant

irradiation: exposure of an article to any form of radiation from infra-red to ultra-violet and including radiations from radio and atomic sources

isotope: one of two or more species of atoms which have essentially the same chemical properties but different atomic mass

LD_{50}: the concentration of a substance required to kill 50 per cent of a group of test animals to which it is administered (Lethal Dose for 50 per cent)

leavening agent: a chemical compound or mixture of compounds which react in doughs to produce carbon dioxide

lecithin: the name given to a group of phosphatides found in cells, particularly egg yolk. Lecithin is a good emulsifier

metabolism: a word describing all the chemical changes which occur in a living organism, plant or animal. Compounds formed during these changes are called *metabolites* and the mechanism, often very complex, by which the changes occur are called *metabolic pathways*

methaemoglobinaemia: methaemoglobin in the blood; methaemoglobin is haemoglobin from decomposing blood

microbe, micro-organism: essentially a living organism of such a size as to be invisible other than through some form of microscope

modifying agent: a compound added to food to influence the texture, particularly of the final product

molecule: the smallest particle of a compound which retains the chemical properties associated with that compound

monomer: the small molecule from which a polymer (see below) is formed

mutagen: a substance which causes or promotes *mutation,* i.e. a permanent chromosomal change in a living cell

mycotoxin: a toxin formed in a food by the growth on it of a particular mould

no-effect level, also No observable effect level (NOEL): the highest concentration of a substance which may be administered to a test animal in any way without causing that animal to be distinguishable from a control to which the substance is not administered

nucleoside: a compound of ribose or desoxy-ribose with one of the purine bases organic: in chemistry, organic compounds are compounds of carbon especially with hydrogen, oxygen and nitrogen but, in effect, with any other element as well

pasteurization: a mild heat treatment designed to kill pathogenic organisms, originally, 145-150°F (63-66°C) for 30 minutes followed by rapid cooling. The modern 'high temperature short time' method of 161°F (72°C) for 15 seconds is now frequently used

pathogen: an organism capable of causing disease in man or animals

pH: a measure of the acidity or alkalinity of a solution or biological system. The pH of pure water is 7 on a scale of 0-14; solutions of pH below 7 are acid and those above 7 are alkaline

phosphatides: a group of compounds closely related to the fats but containing an atom of phosphorus

phospholipids: fats containing phosphorus, e.g. lecithin

poison: that which kills or injures health when ingested or otherwise administered. It has the connotation of rapid action

polyhydric: a substance containing many hydroxyl groups is said to be 'polyhydric'

polymer: a large molecule built up from a number of identical small ones

precursor: a compound which is converted in the body to a desired substance, e.g. a vitamin. The term may also be applied to substances from which *un*desirable substances are formed

protein efficiency ratio (PER): the ratio of the weight gained by an animal to the weight of protein eaten

pyrolysis: the decomposition of substances by high temperatures

sequestrants: substances which have the property of isolating ('sequestrating') other substances, usually metals, from solution and hence preventing them from entering into unwanted reactions

silicones: organic compounds containing silicon and forming polymers of high stability and special properties

sterilization: a process by which all life in a medium or on an article is killed

surface tension: the force at the surface of a liquid tending to reduce the area of it to a minimum and making it behave as an elastic skin

surfactant: a substance which reduces surface tension in a liquid, thus promoting mixing and emulsification

synergism: the combined action of two components such that the result of the two acting together is greater than the sum of the two acting separately. That which increases the effect of another is called a *synergist*

syn. = synonym: used in the naming of organisms to indicate two names for the one species, the first usually superseding the second

teratogen: a compound which, if ingested by a pregnant female, is capable of producing anatomical and physiological defects in the offspring

thiaminase: an enzyme capable of destroying thiamin (vitamin B_1)

thyrotoxicosis: excessive thyroid activity evident as nervousness, loss of weight and increased pulse rate

toxic: poisonous

toxicant: poison

toxin: a poison of biological origin

trihalomethane: a compound of methane (CH_4) in which three of the hydrogen atoms have been substituted by a halogen, i.e. chlorine, iodine, bromine or fluorine

water activity: a property of solutions determined by the number of molecules and/or ions of solute. It may be expressed arithmetically and related to the growth of micro-organisms in foods which may then be formulated so as to ensure microbiological stability

References

Chapter 1

Burnett, J. 'The Adulteration of Foods Act, 1860', *Food Manufacture*, vol. 35, 1960, p. 479.

Drummond, J. C. and A. Wilbraham. *The Englishman's Food*. London, Jonathan Cape, 2nd ed., 1957.

Farrer, K. T. H. *A Settlement Amply Supplied: Food Technology in Nineteenth Century Australia*. Melbourne, M.U.P., 1980, Chapter 12.

Tannahill, R. *Food in History*, London, Methuen, 1973.

Wilson, C. Anne, *Food and Drink in Britain*. London, Constable, 1973.

Chapter 2

Ballester, D., E. Yanez, R. Garcia, S. Erazo, F. Lopez, E. Haardt, S. Cornejo, A. Lopez, J. Pokniak and C. O. Chichester. 'Chemical Composition, Nutritive Value and Toxicological Evaluation of Two Species of Sweet Lupine (*Lupinus albus* and *Lupinus luteus*)', *Journal of Agricultural and Food Chemistry*, vol. 28, 1980, pp 402-5.

Bickel, L. *This Accursed Land*. Melbourne, Macmillan, 1977.

Bressler, R. 'The Unripe Akee – Forbidden Fruit', *Nutrition Reviews*, vol. 34, no. 11, 1976, pp 349-50.

Culvenor, C. C. J., J. A. Edgar and L. W. Smith, 'Pyrrolizidine Alkaloids in Honey from *Echium plantagineum L.*', *Journal of Agricultural and Food Chemistry*, vol. 29, 1981, p. 958.

Gladstones, J. S. 'Lupins as Crop Plants', *Field Crop Abstracts*, vol. 23, no. 2, 1970, pp 123-48.

Gladstones, J. S. 'The Narrow-leafed Lupin in Western Australia', *Western Australia Department of Agriculture Bulletin 3990*, 1977.

Graham, F. M. 'Caffeine – Its Identity, Dietary Sources, Intake and Biological Effects', *Nutrition Reviews,* vol. 36, 1978, pp 97-102.

Hetzel, B. S. and A. J. McMichael. 'Alimentary Tract Cancer in Australia in Relation to Diet and Alcohol', *Nutrition and Cancer,* vol. 1, 1979, pp 82-9.

Hirayama, T. 'Diet and Cancer', ibid., pp 67-81.

Hölscher, P. M. and J. Natzschka. 'Methämoglobinämie bei jungen Säuglingen durch nitrithaltigen Spinat', *Deutsche Medicalische Wochenschrift,* vol. 89, 1964, p. 1751.

Hudson, B. J. F. 'Removal of Toxic Components from Food', *Proceedings of the Institute of Food Science and Technology,* vol. 13, 1980, pp 263-73.

Keeney, D. R. 'Nitrates in Plants and Waters', *Journal of Milk and Food Technology,* vol. 33, 1970, p. 425.

Liener, I. E. ed. *Toxic Constituents of Plant Foodstuffs.* 2nd ed., London, Academic Press, 1980, passim.

Maduagwu, E. N. and D. H. E. Oben. 'Effects of Processing Grated Cassava Roots by the "Screw Press" and by Traditional Fermentation Methods on the Cyanide Content of Gari', *Journal of Food Technology,* vol. 16, 1981, pp 299-302.

Mayo, N. S. 'Cattle Poisoning by Potassium Nitrate', *Kansas Agricultural Experiment Station Bulletin 49,* 1895.

Miller, J. A. and E. C. Miller, 'Carcinogens Occurring Naturally in Foods', *Federation Proceedings,* vol. 35, 1976, pp 1316-21.

Miller, S. A. 'Balancing the Risks Regarding the Use of Nitrites in Meats', *Food Technology,* vol. 34, 1980, pp 254-7.

National Academy of Science (USA). *Toxicants Occurring Naturally in Foods.* 2nd ed., Washington, National Academy of Science, 1973.

Palmer-Jones, T. 'Poisonous Honeys Overseas and in New Zealand', *New Zealand Medical Journal,* vol. 64, 1965, pp 631-37. (See also the series of papers by Palmer-Jones and others in the *New Zealand Journal of Science and Technology,* vol. 29, 1947, pp 107-143.)

Randolph, T. G. *Human Ecology and Susceptibility to the Chemical Environment.* Charles C. Thomas, Springfield, Ill., 1962.

Roberts, H. H. *Food Safety.* New York, John Wiley and Sons, 1981.

Talburt, W. F. and O. Smith, *Potato Processing.* 2nd ed., Westport, Conn., AVI Publishing Co., 1967.

Timson, J. 'Caffeine', *Mutation Research,* vol. 47, 1977, pp 1-52.

Chapter 3

Anon. 'Lead in Food', *British Food Journal,* vol. 78, 1976, p. 72.

Anon. 'DDT May be Good for People', *Nature,* vol. 233, 1977, pp 437-8.

Australian Academy of Science. *Report No. 14, The Use of DDT in Australia.* Canberra, 1972.

Australian Academy of Science. *Report No. 22, Food Quality in Australia.* Canberra, 1977.

Australian Academy of Science. *Health and Environmental Lead in Australia.* Canberra, 1981.

Bebbington, G. N., N. J. Mackay, R. Chvojka, R. J. Williams, A. Dunn and E. H. Auty. 'Heavy Metals, Selenium and Arsenic in Nine Species of Australian Commercial Fish', *Australian Journal of Marine and Freshwater Research,* vol. 28, 1977, pp 277-86.

Bloom, H. *Heavy Metals in the Derwent Estuary*. Hobart, Chemistry Department, University of Tasmania, 1975.

Burden, F. R. Monash University, private communication, 1982.

Chichester, C. O. ed. *Research in Pesticides*. New York, Academic Press, 1965.

Commonwealth Scientific and Industrial Research Organization. *Institute of Industrial Technology, Annual Report*. 1980/81.

Cooper, R. M. 'Polychlorinated Biphenyls – Of Risks and Benefits', *Food Drug and Cosmetic Law Journal,* vol. 39, 1984, pp 240-267.

Department of Health and Social Services, *Lead and Health. The Report of the DHSS Working Party on Lead in the Environment*. London, HMSO, 1973.

Di Ferrante, E. ed. *Trace Metals: Exposure and Health Effects*. London, Pergamon Press, 1979.

Frazier, J. M. *Environmental Health Perspectives,* vol. 28, 1979, p. 39.

Geisman, J. R. 'Reduction of Pesticide Residues in Food Crops by Processing', *Residue Reviews,* vol. 54, 1975, pp 43-53.

Heichel, G. H., L. Hankin and R. A. Botsford. 'Lead in Paper: a Potential Source of Food Contamination', *Journal of Milk and Food Technology,* vol. 37, 1974, pp 499-503.

Johnson, P. E. 'Misuse in Foods of Useful Chemicals', *Nutrition Reviews,* vol. 35, 1977, pp 225-8.

Mackay, N. J., R. J. Williams, J. L. Kacprzac, M. N. Kazacos, A. J. Collins and E. H. Auty. 'Heavy Metals in Cultivated Oysters (*Crassostrea commercialis = Saccostrea cucullata*) from the Estuaries of New South Wales', *Australian Journal of Marine and Freshwater Research,* vol. 26, 1975, pp 31-46.

McKenzie, Joan M. 'Toxic Trace Elements in New Zealand' in *Proceedings of New Zealand Workshop on Trace Elements in New Zealand*. University of Otago, 1981.

Ministry of Agriculture Fisheries and Food, Working Party on the Monitoring of Foodstuffs for Heavy Metals. *Fourth Report. Survey of Cadmium in Food.* London, HMSO, 1973.

Ministry of Agriculture Fisheries and Food, Working Party on the Monitoring of Foodstuffs for Heavy Metals. *Fifth Report. Survey of Lead in Food: First Supplementary Report*. London, HMSO, 1975.

Ministry of Agriculture, Fisheries and Food Study Group on Food Surveillance. *The Surveillance of Food Contamination in the United Kingdom, Food Surveillance Paper No. 1*. London, HMSO, 1978.

Richardson, B. J. and J. S. Waid, 'Polychlorinated Biphenyls (PCBs): An Australian Viewpoint on a Global Problem', *Search,* vol. 13, 1975, pp 17-25.

Roberts, H. R. ed. *Food Safety*. New York, John Wiley and Sons, 1981.

Smith, W. E. and A. M. Smith, *Minimata*. London, Chatto and Windus, 1975.

Thrower, S. J. and I. J. Eustace. 'Heavy Metal Accumulation in Oysters Grown in Tasmanian Waters', *Food Technology in Australia,* vol. 25, 1973, pp 546-52.

Working Group on Mercury in Fish, Department of Primary Industry. *Report on Mercury in Fish and Fish Products*. Canberra, 1980.

Yannai, S. 'Toxic Factors Induced by Processing' in Liener, I. E. ed. *Toxic Constituents of Plant Foodstuffs*. 2nd ed. London, Academic Press, 1980.

Chapter 4

American Council on Science and Health, *Report on Saccharin*. 2nd ed., 1982.

Anon. 'The Cyclamate Bandwagon', *Nature*, vol. 224, 1969, pp 298-9; 'How McPerson's Rule Sank Cyclamates', ibid., pp 398-9; 'How to ban Chemicals without Scaring People', ibid., vol. 225, 1970, pp 3-4; 'Consultation to Curb Panic', ibid., p. 676.

Archer, W. H. *Abstracts of English and Colonial Patent Specifications Relating to the Preservation of Food, etc.* Melbourne, John Ferres, Government Printer, 1870, p. 42, 43, 44, patents granted to I. Leys, R. A. Brooman and J. Gamgee.

Ashby, J., J. A. Styles, D. Anderson and D. Paton, 'Saccharin: An Epigenetic Carcinogen/Mutagen?', *Food and Cosmetic Toxicology*, vol. 16, 1978, pp 95-103.

Baker, G. J., P. Collett and D. H. Allen. 'Bronchospasm Induced by Metabisulphite Containing Foods and Drugs', *Medical Journal of Australia*, 1981, vol. 2, pp 614-16.

Bender, A. E. *Dictionary of Nutrition and Food Technology*. London, Butterworths, 4th ed., 1975.

British Nutrition Foundation, *Food Additives – Why?* London, Forbes Publications, 1977.

Codex Alimentarius Commission Recommended International General Standard for Irradiated Foods and Recommended Code of Practice for the Operation of Radiation Facilities for the Treatment of Foods. CAC/RS 106 – 1979, CAC/RCP 19 – 1979.

Drummond, J. C. and A. Wilbraham. *The Englishman's Food*. London, Jonathan Cape, 2nd ed., 1957.

Elias, P. S. 'Developments in Food Preservation by Ionizing Radiations', *Proceedings of the Institute of Food Science and Technology (UK)*, vol. 15, 1982, pp 71-9.

Farrer, K. T. H. 'Food Additives – For and Against', *Food Technology in Australia*, vol. 27, 1975, pp 379-87.

Farrer, K. T. H. 'The Importance of Salt in Food Technology', *Food and Nutrition Notes and Reviews*, vol. 34, 1977, pp 177-81.

Farrer, K. T. H. 'Is Our Food Safe?' in *Food Quality in Australia*, Australian Academy of Science, Report No. 22, 1977.

Farrer, K. T. H. *A Settlement Amply Supplied: Food Technology in Nineteenth Century Australia*. Melbourne, M.U.P., 1980, Chapter 12.

Heins, H. G. 'Food Irradiation: Dutch Experiences with Practical Applications and Preservative Status in the Netherlands', *Proceedings of the Institute of Food Science and Technology (UK)*, vol. 15, 1982, pp 80-3.

Institute of Food Technologists Expert Panel on Food Safety and Nutrition and the Committee on Public Information. 'Monosodium Glutamate (MSG)', *Food Technology*, vol. 34, 1980, pp 49-63.

International Atomic Energy Agency. 'International Acceptance of Irradiated Food, Legal Aspects'. STI/PUB/530, Vienna, 1979.

Johnson, A. H. and M. S. Peterson. *Encyclopedia of Food Technology*. Westport, Conn., AVI Publishing Co., 1974.

Lueck, E. *Antimicrobial Food Additives: Characteristics, Uses, Effects*. Berlin. Springer Verlag, 1980, pp 115-31.

National Health and Medical Research Council, Canberra. *Approved Food Standards and Approved Food Additives*.

New Zealand Food and Drug Regulations.

Peterson, M. S. and A. H. Johnson, *Encyclopedia of Food Science*. Westport, Conn., AVI Publishing Co., 1978.

Report of the Committee of Enquiry into the Fluoridation of Victorian Water Supplies for 1979-80. Melbourne, Government Printer, 1980.

Sofos, J. N., F. F. Busta and C. E. Allen, 'Botulism Control by Nitrite and Sorbate in Cured Meats: A Review', *Journal of Food Protection*, vol. 42, 1979, pp 739-70.

Tannahill, R. *Food in History*. London, Methuen, 1973.

Tompkin, R. B. and others. An Overview Symposium entitled 'An Assessment of Nitrite for the Prevention of Botulism', *Food Technology*, vol. 34, 1980, pp 229-86.

Weurman, C. and S. van Straten. *Lists of Volatile Flavour Compounds*. 2nd ed. Report No. R 1687, Central Instituut voor Voedingsonderzoek TNO Nederland.

WHO, *Report of a Joint FAO/IAEA/WHO Expert Committee on the Wholesomeness of Irradiated Food. WHO Technical Report Series No. 659*.

Wills, P. A. 'The Use of Atomic Energy for the Irradiation of Foods', *Food Technology in Australia*, vol. 34, 1982, pp 420-4.

Wilson, C. Anne. *Food and Drink in Britain*. London, Constable, 1973.

Chapter 5

Anderson, D. and E. Longstaff. 'An Appraisal of Mutagenicity Test Systems', *The Analyst*, vol. 106, 1981, pp 1-22.

Anon. 'Wodicka Rates Food Additive Hazard as Low', *Food Chemical News*, vol. 12, no. 49, 1971, p. 7.

Anon, 'Experts Find No Link Between Hyperactivity and Diet', *Food Technology in Australia*, vol. 34, 1982, p. 75.

Bonin, A. M. and R. S. U. Baker. 'Mutagenicity Testing of Some Approved Food Additives with the *Salmonella* Microsome Assay', *Food Technology in Australia*, vol. 32, 1980, pp 608-11.

Claus, G., I. Krisko and K. Bolander. 'Chemical Carcinogens in the Environment and in the Human Diet: Can a Threshold be Established?', *Food and Cosmetics Toxicology*, vol. 12, 1974, pp 737-46.

Doll, R. 'Nutrition and Cancer: A Review', *Nutrition and Cancer*, vol. 1, 1979, pp 35-45.

Ebert, A. G. 'We Cannot Prove Safety', *Food Product Development*, vol. 11, 1977, pp 26-33.

Farrer, K. T. H. 'Is Our Food Safe?' in *Food Quality in Australia*, Australian Academy of Science, Report No. 22, 1977.

Gehring, P. 'The Risk Equations: The Threshold Controversy', *New Scientist,* 18 August 1977, pp 426-8.

Goldberg, L. 'Safety Evaluation Concepts', *Journal of the Association of Official Agricultural Chemists,* vol. 58, 1975, pp 635-44.

Goldberg, L. 'The Safe and Effective Issue 2 – A Scientist's View', *Research Management,* vol. 19, 1976, pp 10-13.

Hathcock, J. N. 'Nutrition: Toxicology and Pharmacology', *Nutrition Reviews,* vol. 34, 1976, pp 65-9.

Kolbye, A. C. 'Short-term Mutagenicity Tests Eyed with Caution', *Food Chemical News,* 29 June 1981, p. 24.

Kroes, R., G. J. van Esch and J. W. Weiss, 'Philosophy of "No-effect" Level for Carcinogens', *Proceedings of the International Symposium on Nitrite Meat Products, Zeist, 1973,* Pudoe, Wageningen, pp 227-41.

Nutrition Foundation. *The National Advisory Committee on Hyperkinesis and Food Additives: Final Report of the Nutrition Foundation.* Washington, D.C., The Nutrition Foundation, 1980.

Oser, B. L. 'Benefit/Risk: Whose? What? How Much?', *Food Technology,* vol. 32, no. 8, 1978, pp 55-8.

Purchase, I. F. H., E. Longstaff, J. Ashby, J. A. Styles, D. Anderson, P. A. Lefevre and F. R. Westwood. 'An Evaluation of 6 Short-term Tests for Detecting Organic Chemical Carcinogens', *British Journal of Cancer,* vol. 37, 1978, pp 873-923.

Randolph, T. C. *Human Ecology and Susceptibility to the Chemical Environment.* Charles C. Thomas, Springfield, Ill., 1962.

Roberts, H. R. *Food Safety.* New York, John Wiley and Sons, 1981.

Schmidt, A. M. 'Food and Drug Law in the United States: a 200 Year Perspective' in *Food Quality and Safety: a Century of Progress.* London, HMSO, 1976.

Sherman, H. 'Toxicology, A Discipline under Attack', *Chemistry in Australia,* vol. 47, 1980, pp 446-52.

Social and Economic Committee of the Food Safety Council (USA). 'Principles and Processes for Making Food Safety Decisions', *Food Technology,* vol. 34, 1980, pp 79-125.

Taylor, S. L. 'Caution in Using Ames Test Advised', *Food Chemistry News,* 29 June 1981, p. 25.

Tepper, L. B. 'The Safe and Effective Issue 1 – A Regulator's View, *Research Management,* vol. 19, 1976, pp 7-9.

Truswell, A. S., N-G. Asp, W. P. T. James and B. MacMahon. 'Food and Cancer. Special Report', *Nutrition Reviews,* vol. 36, 1978, pp 313-14.

Upton, Arthur C. 'The Biological Effects of Low-Level Ionizing Radiation', *Scientific American,* February 1982.

Vettorazzi, G. "The Safety Evaluation of Food Additives: the Dynamics of Toxicological Decision', *Lebensmittel-Wissenschaft und Technologie,* vol. 8, 1975, pp 195-201 (in English).

Wattenberg, L. W., W. D. Loub, L. K. Lam and J. L. Speier. 'Dietary Constituents Altering the Responses to Chemical Carcinogens', *Federation Proceedings,* vol. 35, 1976, pp 1327-31.

WHO. 'Evaluation of Certain Food Additives', *WHO Technical Report Series 669,* 1981.

Chapter 6

Bauman, H. E. 'The HACCP Concept and Microbiological Hazard Categories', *Food Technology*, vol. 28, 1974, pp 30-4, 74.

Bigalke, D. L. and F. F. Busta. 'Quality Management Systems for the Food Industry. HACCP Approach to Improved Profits', *Food Technology in Australia*, vol. 34, 1982, pp 515-17.

Carefoot, G. L. and E. R. Sprott, *Famine on the Wind: Plant Diseases and Human History*. Sydney, Angus and Robertson, paperback ed., 1974, Chapter 1.

Christian, J. H. B. 'Developments in Microbiological Criteria for Foods', *Food Technology in Australia*, vol. 34, 1982, pp 498-9.

CSIRO Consumer Service. 'How to Handle Chicken – Fresh or Frozen' (a consumer leaflet available on request).

Eisenberg, M. S., K. Gaarslev, W. Brown, M. Horwitz and D. Hill. 'Staphylococcal Food Poisoning Aboard a Commercial Aircraft', *Lancet*, 27 September 1975, pp 595-9.

Gilbert, J., M. J. Shepherd, M. A. Wallwork and M. E. Knowles, 'A Survey of the Occurrence of Aflatoxin M_1 in UK-produced Milk for the Period 1981-1983', *Food Additives and Contaminants*, vol. 1, 1984, pp 23-28.

Hobbs, B. C. and R. J. Gilbert, *Food Poisoning and Food Hygiene*. London, Edward Arnold, 4th ed., 1978.

Hobbs, B. C. 'Observations on Public Health Microbiology Including Experience in India', *Food Technology in Australia*, vol. 34, 1982, pp 501-7.

Howie, J. 'Policies in the U.K. to Ensure that a Food Factory does not Distribute Food-Poisoning Micro-Organisms – A Personal View', *Journal of Applied Bacteriology*, vol. 47, 1979, pp 233-5.

Kiermeier, F. 'The Significance of Aflatoxins in the Dairy Industry', *International Dairy Federation. Doc. No. 98*, 1977.

Merson, M. H., J. M. Hughes, D. N. Lawrence, J. G. Wells, J. J. D'Agnese and J. C. Yashuk, 'Food and Waterbone Disease Outbreaks on Passenger Cruise Vessels and Aircraft', *Journal of Milk and Food Technology*, vol. 39, 1976, pp 285-8.

Moss, M. O. 'Role of Mycotoxins in Disease of Man', *International Journal of Environmental Studies*, vol. 8, 1975, pp 165-70.

Osborne, B. G. 'Mycotoxins and the Cereal Industry – A Review, *Journal of Food Technology*, vol. 17, 1982, pp 1-9.

Pitt, J. I. 'Mycotoxins in Human Health', *CSIRO Food Research Quarterly*, vol. 41, 1981, pp 31-7.

Rodericks, J. V. ed. *Mycotoxins and Other Fungal Related Food Problems*. Washington, D.C., American Chemical Society, 1976.

Chapter 7

FAO, Food Additive Control Series, Rome, FAO, 159-1963, No. 1 Canada, No. 2 United Kingdom, No. 3 The Netherlands, No. 4 Australia, No. 5 Denmark, No. 6 France, No. 7 Federal Republic of Germany.

Farrer, K. T. H. *A Settlement Amply Supplied: Food Technology in Nineteenth Century Australia*. Melbourne, M.U.P., 1980, Chapter 12.

Hall, R. L. 'GRAS Review and Food Additive Legislation', *Food Technology*, vol. 25, 1971, pp 466-70.

MAFF. *Food Quality and Safety: a Century of Progress*. London, HMSO, 1976. Includes papers by A. M. Schmidt on United States food laws and by G. O. Kermode on developing countries.

Moss, M. O. 'Role of Mycotoxins in Disease of Man', *International Journal of Environmental Studies*, vol. 8, 1975, pp 165-70.

Turner, A. 'Food and Public Safety', *Proceedings of the Institute of Food Science and Technology (UK)*, vol. 13, 1980, pp 235-50.

Index